T0173432

Cambridge Lower Secondary

Maths

STAGE 8: STUDENT'S BOOK

Alastair Duncombe, Belle Cottingham, Rob Ellis, Amanda George, Brian Speed

Series Editor: Alastair Duncombe

Collins

William Collins' dream of knowledge for all began with the publication of his first book in 1819.

A self-educated mill worker, he not only enriched millions of lives, but also founded a flourishing publishing house. Today, staying true to this spirit, Collins books are packed with inspiration, innovation and practical expertise. They place you at the centre of a world of possibility and give you exactly what you need to explore it.

Collins. Freedom to teach.

Published by Collins
An imprint of HarperCollins*Publishers*
The News Building
1 London Bridge Street
London
SE1 9GF

HarperCollins*Publishers*
Macken House, 39/40 Mayor Street Upper,
Dublin 1, D01 C9W8, Ireland

> Browse the complete Collins catalogue at
> **www.collins.co.uk**

MIX
Paper | Supporting
responsible forestry
FSC™ C007454

This book is produced from independently certified FSC™ paper to ensure responsible forest management.

For more information visit:
www.harpercollins.co.uk/green

British Library Cataloguing in Publication Data
A catalogue record for this publication is available from the British Library.

Authors: Alastair Duncombe, Belle Cottingham, Rob Ellis, Amanda George, Brian Speed
Series editor: Alastair Duncombe
Publisher: Elaine Higgleton
In-house project editors: Jennifer Hall and Caroline Green
Project manager: Wendy Alderton
Development editors: Phil Gallagher and Jess White
Copyeditor: Alison Bewsher
Proofreader: Eric Pradel
Answer checkers: Eric Pradel and Jouve India Private Limited
Cover designer: Ken Vail Graphic Design and Gordon MacGlip
Cover illustrator: Ann Paganuzzi
Typesetter: Jouve India Private Limited

Production controller: Lyndsey Rogers
Printed and bound in India by Replika Press Pvt. Ltd.

Acknowledgements

The publishers gratefully acknowledge the permission granted to reproduce the copyright material in this book. Every effort has been made to trace copyright holders and to obtain their permission for the use of copyright material. The publishers will gladly receive any information enabling them to rectify any error or omission at the first opportunity.

p.45 tommyboy1289/Shutterstock; p.88 Hannah Ritchie (2018) - "Urbanization". Published online at OurWorldInData.org. Retrieved from: 'https://ourworldindata.org/urbanization' [Online Resource] Data available under Creative Commons Attribution 4.0 International (CC BY 4.0); p.93 © December 2019 by PopulationPyramid.net, made available under a Creative Commons license CC BY 3.0 IGO: http://creativecommons.org/licenses/by/3.0/igo/; p. 98 UNdata http://data.un.org; p.102 International Paralympic Committee, https://www.paralympic.org; p.103 Japan Meteorological Agency website (https://www.data.jma.go.jp/obd/stats/etrn/view/monthly_s3_en.php?block_no=47412&view=1) p.197 Hannah Ritchie (2019) – "Age Structure". Published online at OurWorldInData.org. Retrieved from: 'https://ourworldindata.org/age-structure' [Online Resource] Data available under Creative Commons Attribution 4.0 International (CC BY 4.0); p. 201 FAO. 2011. The state of the world's land and water resources for food and agriculture (SOLAW) – Managing systems at risk. Food and Agriculture Organization of the United Nations, Rome and Earthscan, London; p.202 © December 2019 by PopulationPyramid.net, made available under a Creative Commons license CC BY 3.0 IGO: http://creativecommons.org/licenses/by/3.0/igo/; p.203 WWF South Africa https://wwf.panda.org/wwf_news/?293410/South%5FAfrica%5Frhino%5Fpoaching%5Ffigures%5F2016; p. 203 United Nations, Department of Economic and Social Affairs, Population Division (2019). World Population Ageing 2019: Highlights (ST/ESA/SER.A/430). Copyright © 2019 by United Nations, made available under a Creative Commons license (CC BY 3.0 IGO); p.206 © FAO 2010, Annual Change in Forest Region by region, 1990-2010, http://www.fao.org/forestry/fra/69406/en/ 16/09/2020; p. 214 Hannah Ritchie (2018) – "FAQs on Plastics". Published online at OurWorldInData.org. Retrieved from: 'https://ourworldindata.org/faq-on-plastics' [Online Resource]. Data available under Creative Commons Attribution 4.0 International (CC BY 4.0); p.217 © December 2019 by PopulationPyramid.net, made available under a Creative Commons license CC BY 3.0 IGO: http://creativecommons.org/licenses/by/3.0/igo/; p.314 [© FAO] 2020 FAOSTAT Land Use data http://www.fao.org/faostat/en/#data/RL 15/09/2020; p. 319 International Olympic Committe: The Olympic Games and all Olympic Games data used in this publication are the property of the International Olympic Committee – All Rights Reserved. Used under licence; p. 319 'World Population Prospects: Key findings & advance tables 2015 revision, by the Department of Economic & Social affairs, © 2015 United Nations. Used with the permission of the United Nations.

Cambridge International copyright material in this publication is reproduced under licence and remains the intellectual property of Cambridge Assessment International Education.

Third-party websites and resources referred to in this publication have not been endorsed by Cambridge Assessment International Education.

With thanks to the following contributors from the first edition: Michele Conway, Sarah Sharratt, Mike Fawcett, Caroline Fawcus, Lisa Greenstein, Deborah McCarthy and Fiona Smith.

With thanks to the following teachers and schools for reviewing materials in development: Samitava Mukherjee and Debjani Sen, Calcutta International School; Hawar International School; Adrienne Leisztinger, International School of Budapest; Sujatha Raghavan, Manthan International School; Mahesh Punjabi, Podar International School; Taman Rama Intercultural School; Utpal Sanghvi International School.

Introduction

The *Collins Lower Secondary Maths Stage 8 Student's Book* covers the Cambridge Lower Secondary Mathematics curriculum framework (0862). The content has been covered in 25 chapters. The series is designed to illustrate concepts and provide practice questions at a range of difficulties to allow you to build confidence on a topic.

The authors have included plenty of worked examples in every chapter. These worked examples will lead you, step-by-step, through the new concepts. They include clear and detailed explanations. Where possible, links have been made between topics, encouraging you to build on what you know already, and to practise mathematical concepts in a different context.

Each chapter within the book contains activities and questions to help you to develop your skills with *thinking and working mathematically*. You will practise the skills of *specialising* and *generalising*, *conjecturing* and *convincing*, *characterising* and *classifying*, and *critiquing* and *improving*. You can find definitions of these characteristics on the next page. The activities and questions will help you to understand each topic. They will also develop your skills at spotting patterns and solving mathematical problems. These activities and questions are indicated by a star:

Every chapter has these helpful features:
- 'Starting point': to remind you of what you know already

- 'This will also be helpful when …': to let you know where you will use the mathematics in the future

- 'Getting started': to get you interested in the new topic through an activity

- 'Key terms' boxes: to identify new mathematical words you need to know in that chapter, and provide a definition

- Clear topic headings: so that you can see what you are going to be learning in each section of the chapter

- Worked examples: to show you how to answer questions with both formal and informal (diagrammatic) explanations provided

- 'Tip' boxes: to give you guidance on the possible methods and common errors

- Exercises: to give you practice at answering questions on each topic. The questions at the end of each exercise will be harder, in order to stretch you

- 'Thinking and working mathematically' questions and activities (marked as ▼): to help you develop your mathematical thinking. The activities will often be more open-ended in nature

- 'Think about' boxes: to suggest ideas that you might want to consider

- 'Discuss' boxes: to encourage you to talk about mathematical ideas with a partner or in class

- 'Did you know?' boxes: to explain where mathematical ideas came from and how they are applied in real life.

- 'Consolidation exercise': to give you further practice on all the topics introduced in the chapter

- 'End of chapter reflection': to help you think about how well you have understood the ideas in the chapter, so that you can monitor your own progress.

We hope that you find this approach enjoyable and engaging as you progress through your mathematical journey.

The Thinking and Working Mathematically Star

Critiquing

Comparing and evaluating mathematical ideas, representations or solutions to identify advantages and disadvantages.
For example:

- Which is the best way to …?
- Write down the advantages and disadvantages of …

Specialising

Choosing an example and checking to see if it satisfies or does not satisfy specific mathematical criteria.
For example:

- Find an example of …
- Find … if … and …

Conjecturing

Forming mathematical questions or ideas.
For example:

- What would happen if …?
- How would you …?

Convincing

Presenting evidence to justify or challenge a mathematical idea or solution.
For example:

- Prove that …
- Explain why …
- Show that …

Improving

Refining mathematical ideas or representations to develop a more effective approach or solution.
For example:

- Find a better way to …
- Describe a more efficient way to …

Classifying

Organising objects into groups according to their mathematical properties.
For example:

- Match …
- Sort …
- Put a ring around all the … which …

Generalising

Recognising an underlying pattern by identifying many examples that satisfy the same mathematical criteria.
For example:

- Find a rule that connects … and …
- What can you conclude from …?

Characterising

Identifying and describing the mathematical properties of an object.
For example:

- What do … have in common?
- Describe the properties of …

Specialising and Generalising

Conjecturing and Convincing

Critiquing and Improving

Characterising and Classifying

Contents

Negative numbers, indices and roots

You will learn how to:

- Estimate, multiply and divide integers, recognising generalisations.
- Recognise squares of negative and positive numbers, and corresponding square roots.
- Recognise positive and negative cube numbers, and the corresponding cube roots.
- Use positive and zero indices, and the index laws for multiplication and division.

Starting point

Do you remember…

- how to estimate the results of multiplications and divisions?

  For example, estimate 56×181

- how to multiply and divide integers if one integer is negative?

  For example, find -7×6, $-32 \div 4$ and $18 \div (-3)$

- how squares and square roots are related?

  For example, write the value of $\left(\sqrt{5}\right)^2$

- how cubes and cube roots are related?

  For example, if $9^3 = 729$, what is the cube root of 729?

- the meanings of positive integer powers?

  For example, write 10^5 as an ordinary number.

This will also be helpful when…

- you learn about surds (square roots or cube roots that are not integers)
- you learn to estimate surds
- you learn about negative indices, and use them in multiplication and division.

1.0 Getting started

To find 18×15, you can split both 18 and 15 into two parts, and then multiply the parts using a grid method:

×	10	8
10	100	80
5	50	40

So, $18 \times 15 = 100 + 80 + 50 + 40 = 270$

- There are other ways to split 18 and 15. Use a grid method to find 18×15 using $18 = 9 + 9$ and $15 = 6 + 9$. You should still get the answer 270.

- You can write 18 = 20 − 2 and 15 = 10 + 5. Copy the grid below and show how to use it to find 18 × 15.

×	20	−2
10		−20
5		

- You can also write 18 = 20 − 2 and 15 = 20 − 5. Copy the grid below and fill in the three empty squares.

×	20	−2
20		
−5		?

- Use the fact that 18 × 15 = 270 to work out the missing number in the '?' square.

- Choose a different pair of two-digit positive integers. Write each number as two positive parts, and multiply using the grid method.

- Write each of the integers as a positive part and a negative part. Deduce the result of multiplying the two negative parts.

- What type of number is the product of two negative numbers? Explain how you know.

- What do you think is the value of −10 ÷ (−2)? Explain your answer.

1.1 Multiplying and dividing positive and negative integers

Worked example 1

Find:

a) $-4 \times (-5)$ **b)** $-24 \div (-8)$ **c)** $(-3)^2$

a) $-4 \times (-5) = 20$	Multiplication by a negative number is repeated subtraction. $-4 \times (-5)$ is the same as 'subtract −5 four times'. This is the same as adding 20.	You can use these rules for multiplication: + × + = + + × − = − − × + = − − × − = +
b) $-24 \div (-8) = 3$	How many −8s make −24? $-8 \times 3 = -24$, so $-24 \div (-8) = 3$	You can use these rules for division: + ÷ + = + + ÷ − = − − ÷ + = − − ÷ − = + **Think about** Compare the rules for multiplication and division. What do you notice?
c) $(-3)^2 = 9$	The brackets tell you to square the number inside, so $(-3)^2$ means $-3 \times (-3)$. $(-3)^2 = -3 \times (-3) = 9$	

Worked example 2

Estimate the answer to each calculation. Then use a calculator to find the exact answer.

a) $-58 \times (-316)$

b) $-812 \div (-29)$

a) $-58 \times (-316) \approx -60 \times (-300)$ $= 18\,000$	Round so that you can do the multiplication in your head. 58 is closer to 60 than to 50, and 316 is closer to 300 than to 400. 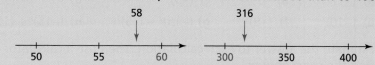
$-58 \times (-316) = 18\,328$	Find the exact answer using your calculator. Use the sign change key usually marked (−) or +/−, to enter a negative number.
b) $-812 \div (-29) \approx -800 \div (-30)$ $= -80 \div (-3) \approx 26$	Round so that you can do the division in your head. 812 is closer to 800 than to 900, and 29 is closer to 30 than to 20. 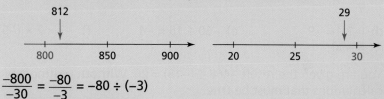 $\dfrac{-800}{-30} = \dfrac{-80}{-3} = -80 \div (-3)$ Ignore any remainder.
$-812 \div (-29) = 28$	Find the exact answer using your calculator.

Exercise 1   1–4, 6–7, 9–12

1 Find:

    a) $-2 \times (-3)$ **b)** $10 \div (-2)$ **c)** $-1 \times (-1)$ **d)** $-15 \div (-3)$

    e) -6×12 **f)** $-56 \div (-8)$ **g)** $-5 \times (-11)$ **h)** $-1 \div 1$

2 Find:

    a) $-3 \times (-2) \times 4$ **b)** $-5 \times 7 \times (-2)$

    c) $-4 \times (-2) \times (-9)$ **d)** $7 \times (-2) \times 3$

3 Write two multiplication calculations and two division calculations that give each answer.

    Include at least one negative number in each calculation.

    a) 16   **b)** 6   **c)** -20

4 Estimate the answer to each calculation.
    Then find the exact answer.

    a) $-88 \times (-47)$ **b)** -37×286

    c) $-228 \div (-12)$ **d)** $-704 \div (-22)$

5 Estimate the answer to each calculation. Then find the exact answer using a calculator.

a) $-4029 \div 79$

b) $-813 \times (-907)$

c) $-6204 \div (-66)$

d) $739 \times (-136)$

6 Vikingur wants to find -581×-779 using his calculator, but the sign change (+/−) key is broken. Explain how he can find the answer without doing written multiplication.

7 Find:

a) $(-7)^2$
b) -6^2
c) $(-3)^3$
d) -2^3

e) $(-10)^3$
f) $(-8)^2$
g) $(-15)^2 - (15)^2$
h) $(-15)^2 + (15)^2$

8 Estimate the answer to each calculation. Then use a calculator to find the exact answer.

a) $-(18)^3$
b) $(-32)^2$
c) $(-101)^3$
d) $-(49)^2$

9 Write true or false for each calculation. If it is false, correct it.

a) $-5 \times (-7) = -35$ 7

b) $(-2)^2 = -4$

c) $-77 \div (-11) = $

d) $(-3)^3 = -9$ $= -30$

e) $-60 \div 15 = -4$

f) $-3 \times 2 \times (-5)$

> **Discuss**
>
> Type each calculation into a calculator exactly: -5^2 and $(-5)^2$. Why are the answers different?
>
> Type each calculation into a calculator exactly: -5^3 and $(-5)^3$. Why are the answers the same?

10 You are given the result $-468 \div (-18) = 26$. Without writing any working, write at least six other calculations that must be true.

> **Think about**
>
> How many different integer calculations can be written without doing any working out? Explain your answer.

11 Is each statement always, sometimes or never true?

If a statement is always true or never true, explain why.

If a statement is sometimes true, give one example where it is true and one example where it is not true.

a) If you multiply two negative integers, the result is greater than both of the numbers.

b) If you multiply a positive integer by a negative integer, the result lies between the two numbers.

c) If you divide a negative integer by a negative integer, the result is greater than both of the numbers.

d) If you divide a positive integer by a negative integer, the result lies between the two numbers.

 **12** Below are two puzzles. In each puzzle, write numbers in the four boxes to make all of the calculations correct.

a)

| | × | | = −30 |

×   ÷

| | × | | = 18 |

= 12   = −5

b)

| | ÷ | | = −9 |

÷   ×

| | ÷ | | = 2 |

= 3   = −6

Thinking and working mathematically activity

Fill in a copy of this multiplication table. First, write the products of two positive numbers, and then write the products of one positive number and one negative number. Use patterns in these results to find the products of two negative numbers.

×	−3	−2	−1	0	1	2	3
3							
2							
1							
0							
−1							
−2							
−3							

Shade positive numbers one colour, negative numbers in a second colour and zero in a third colour.

What patterns can you see in the table?

What does the table tell you about multiplication of two negative numbers?

What does the table tell you about division with positive and negative numbers? Explain your answer.

1.2 Squares, cubes and roots of positive and negative numbers

Worked example 3

a) Find the positive and negative square roots of 400.

b) Find $\sqrt[3]{-8}$

a) The square roots of 400 are 20 and −20.	400 has two square roots because $(20)^2 = 20 \times 20 = 400$ and $(−20)^2 = −20 \times (−20) = 400$.
b) $\sqrt[3]{-8} = −2$	$−2 \times (−2) \times (−2) = −8$, so the cube root of −8 is −2.

Key terms

The **square root** of a number squares to make the number. A number has two square roots: one is positive and the other is negative. For example, the square roots of 9 are 3 and −3.

The **cube root** of a number cubes to make the number. For example, the cube root of 27 is 3 because $3 \times 3 \times 3 = 27$.

1 Write down both square roots of each number.

a) 81   b) 1   c) 10 000   d) 400

2 Find:

a) $\sqrt[3]{-27}$ b) $\sqrt[3]{-125}$ c) $\sqrt[3]{(-8)^2}$ d) $\left(\sqrt[3]{-12}\right)^3$

> **Did you know?**
>
> If a question asks for $\sqrt{4}$, this means the positive square root, 2.
>
> The symbol ± (plus or minus) can be used with a square root sign. If a question asks for $\pm\sqrt{4}$, this means both square roots, 2 and −2.

3 Write true or false for each calculation. If it is false, correct it.

a) $(-5)^2 = 25$ b) $(-2)^3 = 8$ c) $-5^2 = -25$ d) $\sqrt{9} = \pm 3$

e) $\sqrt[3]{1} = \pm 1$ f) $\sqrt{-4} = \pm 2$ g) $(-3)^3 = 27$ h) $\sqrt[3]{-64} = -4$

4 a) Write three integers that do not have a square root.

b) Write two numbers where the cube root of the number equals the number.

5 Use a calculator to find the values.

a) $\sqrt[3]{-512}$ b) $\sqrt[3]{-343} + \sqrt[3]{4096}$ c) $\sqrt{13^2 - 12^2}$ d) $\sqrt[3]{1 - 9^3 - 10^3}$

6 Use a calculator to find the values. Round your answers to two decimal places.

a) $\sqrt[3]{-99}$ b) $\sqrt[3]{4.6} - \sqrt[3]{-3.7}$ c) $\sqrt{8.8^2 - 2.2^2 \times 2}$

d) $\sqrt[3]{5.6^2 - 3}$

> **Tip**
>
> In the order of operations, treat calculations under a root as if they are inside a bracket. For example,
>
> $\sqrt{13^2 - 12^2} = \sqrt{\left(13^2 - 12^2\right)}$

7 **Technology question** What happens when you try to calculate $\sqrt{-100}$ using a calculator? Why?

8 Without finding the values, write these in increasing order.

$\sqrt[3]{-50}$ $\sqrt[3]{50}$ $\sqrt{50}$ $-\sqrt{50}$

9 If $(-2)^{17} = -131\,072$, find the value of $(-2)^{16}$. Show your method.

10 a) Explain why a positive number has two square roots.

b) Explain why a negative number cannot have a positive cube root.

c) How many different square roots does 0 have? Explain your answer.

> **Think about**
>
> $1^2 = 1$
>
> $11^2 = 121$
>
> $111^2 = 12\,321$
>
> Predict 1111^2
>
> What would $111\,111^2$ be?
>
> Use a calculator to check your predictions.

Thinking and working mathematically activity

Part 1

Write the first twenty-one square numbers, starting with 0^2. Look at the final digits of the squares. Describe any pattern you can see. Use your findings to answer the questions below.

Why are there no square numbers ending in 7?

Without doing any calculations, state which of these numbers cannot be square numbers. Explain your answer.

1423   6241   7744   8838

Without doing any calculations, predict the units digit for each of these square numbers.

$(608)^2$ $(753)^2$ $(831)^2$ $(926)^2$

Part 2

Write the values of the cube numbers from 2^3 to 6^3.

Can you write each result as a sum of consecutive odd numbers?

Describe any patterns you can see.

Use your findings to find 13^3 using only addition. (You can use a calculator to do the addition.) Discuss with other students how you did this. What is the most efficient method?

1.3 The index laws

Key terms

A **power** or **index** tells you how many copies of a number to multiply together.
For example, $(10)^5$ means $10 \times 10 \times 10 \times 10 \times 10 = 100\,000$. $(10)^5$ is called 'the fifth power of 10' or '10 to the power of 5' (or '10 to the 5' for short). The plural of index is indices.

Any number to the **power of 0** equals 1.

To **multiply** powers of a number, **add** the powers. For example, $3^3 \times 3^2 = 3^{3+2} = 3^5$

To **divide** powers of numbers, **subtract** the powers. For example, $3^3 \div 3^2 = 3^{3-2} = 3^1 = 3$

To find the **power of a power, multiply** the powers. For example, $(3^3)^2 = 3^{3 \times 2} = 3^6$

Worked example 4

Write each expression as a power of 5.

a) $5^2 \times 5^4$ **b)** $5^5 \div 5^3$ **c)** $(5^4)^2$

a) $5^2 \times 5^4$	$5^2 = 5 \times 5$
$= 5^{2+4}$	$5^4 = 5 \times 5 \times 5 \times 5$
$= 5^6$	Multiply them together.
	$5^2 \times 5^4 = 5 \times 5 \times 5 \times 5 \times 5 \times 5$
	$= 5^6$

b) $5^5 \div 5^3$

$= 5^{5-3}$

$= 5^2$

$5^5 = 5 \times 5 \times 5 \times 5 \times 5$
$5^3 = 5 \times 5 \times 5$

Write the division as a fraction.

$5^5 \div 5^3 = \dfrac{5^5}{5^3}$

Simplify the fraction.

$$\dfrac{5 \times 5 \times \cancel{5} \times \cancel{5} \times \cancel{5}}{\cancel{5} \times \cancel{5} \times \cancel{5}}$$

$$= \dfrac{5 \times 5}{1}$$

$$= 5 \times 5$$

$$= 5^2$$

c) $(5^4)^2 = 5^{4 \times 2}$

$= 5^8$

$(5^4)^2 = 5^4 \times 5^4$

$= (5 \times 5 \times 5 \times 5) \times (5 \times 5 \times 5 \times 5)$

$= 5^8$

Exercise 3

1 Write as a single power:

a) $3^6 \times 3^2$

b) $4^7 \times 4^1$

c) $5^2 \times 5^6$

d) $2^3 \times 2^2 \times 2^2$

e) $6^3 \times 6^0 \times 6^2$

f) $7^5 \times 7^4 \times 7$

2 Copy and complete these number pyramids. The number in each cell is the product of the numbers in the two cells below it.

a)

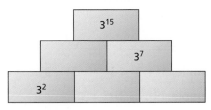

b)

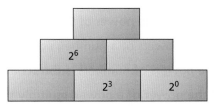

3 Simplify each expression. Leave your answer in index form.

a) $3^6 \div 3^2$

b) $3^8 \div 3^2$

c) $4^{11} \div 4^6$

d) $6^7 \div 6^0$

e) $5^4 \div 5^2 \times 5^3$

f) $7^3 \times 7^4 \div 7^5$

4 Simplify each expression. Leave your answer in index form.

a) $(2^3)^2$

b) $(7^5)^2$

c) $(8^{10})^3$

5 Write the value of:

a) $6^9 \div 6^8$

b) $10^3 \times 10^4$

c) 6.77^0

d) $(2^2)^3$

e) 37^1

f) $4^8 \div 4^6$

6 Simplify each expression. Leave your answer in index form.

a) $\dfrac{3^3 \times 3^1}{3}$

b) $\dfrac{5^3 \times 5^4}{5^2}$

c) $\dfrac{6^{11} \times 6^3}{6^5}$

Tip

Each expression can be rewritten using brackets, e.g. $\dfrac{3^3 \times 3^1}{3} = (3^3 \times 3^1) \div 3$

7 Which of these numbers are not equal to 1 million?

a) $10^4 \times 10^2$

b) $10^2 \times 10^2 \times 10^2$

c) $\dfrac{10^3 \times 10^3}{10}$

d) $10^3 + 10^3$

8 Explain why you cannot use an index law with $2^3 \times 3^2$

9 Lucia thinks that $2^2 \times 2^3 = 4^5$

Do you agree with her? Give reasons for your answer.

10 For the calculation $3^2 \times 3^2 \times 3^2$, Jake wrote $3^{2 \times 2 \times 2} = 3^8$. Explain why Jake is incorrect. Find the correct answer.

11 The table shows some of the powers of 6.

Power	6^0	6^1	6^2	6^3	6^4	6^5	6^6
Value	1	6	36	216	1296	7776	46 656

a) Use the table to explain why $36 \times 1296 = 46\ 656$

b) Find: $7776 \div 36$

12 Write each number as a power of 2 or a power of −2.

a) 8

b) 64

c) 1

d) −128

13 Find the missing numbers.

a) $8 \times 2^4 = 2^{\square}$

b) $2^{\square} \times 64 = 2^{10}$

c) $64 \div \square = 2^3$

Think about

In the table in question 11, why do all the powers of 6 end in 6?

Tip

Some of your answers to question 12 can help you with question 13.

Discuss

What could ★ be if $(\bigstar)^2 \times (\bigstar)^2 = 81$?

Thinking and working mathematically activity

Part 1

Copy and complete the table below. On the right, write the number you divide by to get the next power.

$2^6 =$	$2 \times 2 \times 2 \times 2 \times 2 \times 2$	$= 64$
$2^5 =$		
$2^0 =$		

$\Big) \div$
$\Big) \div$
$\Big) \div$
$\Big) \div$
$\Big) \div$

Use your results to explain why it makes sense that $2^0 = 1$.

Do the powers of other integers follow the same pattern? If you are not sure, create a similar table for powers of 10.

Part 2

For each of the multiplication and division laws for indices, follow the three steps below.

Step 1:

Write an example, making one of the indices 0.

For example, $2^3 \times 2^0$ for the first index law.

Step 2:

Use the index law to rewrite as a single power. Find the value of the answer.

For example, write $2^3 \times 2^0$ as a single power of 2, and work out its value.

Step 3:

Find the value of your example without using an index law. Is it the same?

For example, find the value of $2^3 \times 2^0$ by working out the values of 2^3 and 2^0 and then multiplying.

Using your results, explain whether the index laws work when one of the indices is zero.

Does each index law work if *both* indices are zero? Use examples to explain your answer.

Consolidation exercise

1 **a)** Copy and complete this multiplication grid.

×		5	–3	
		–15	9	
–2				
8	–32			
				20

> **Discuss**
>
> Which columns and rows could have different answers? Which could not? Why?

b) Compare your answers with other students' answers.

2 Write the number that:

a) has cube root −4   b) has square roots −6 and 6   c) has cube −1000

3 Decide whether each statement is always true, sometimes true or never true.

If a statement is always true or never true, explain why.

If a statement is sometimes true, write one example where it is true and one example where it is not true.

a) An integer multiplied by itself makes a positive integer.

b) The cube of an integer is a positive integer.

c) The square of the cube of a negative integer is a positive integer.

4 Find the missing power in each statement.

a) $3^\square \times 3^2 = 3^5$

b) $5^2 \times 5^\square = 5^7$

c) $6^3 \times 6^\square \times 6^5 = 6^{12}$

d) $4^{10} \div 4^\square = 4^5$

e) $8^\square \div 8^2 = 8^3$

f) $\dfrac{7^6 \times 7^3}{7^\square} = 7^3$

g) $(9^\square)^2 = 9^{20}$

h) $(3^6)^\square = 3^0$

5 To find the value of $2^4 \times 2^0$, Bella wrote:

$2^4 \times 2^0 = 2^4 \times 0 = 0.$

Explain why Bella is incorrect. Find the correct answer.

6 For the calculation $2^3 \times 4^2$, Keshini wrote:

$2 \times 4 = 8$ and $3 + 2 = 5$.

So the answer is 8^5.

Explain why Keshini is incorrect. Find the correct answer.

7 a) Write 8 and 128 as powers of 2.

b) Use your answer to complete:

i) $8 \times 128 = 2^\square$ ii) $128 \div 8 = 2^\square$ iii) $8 \times 8 \times 128 = 2^\square$ iv) $\dfrac{2^7 \times 8}{64} = 2^\square$

> **Discuss**
>
> Arrange the numbers 1 to 15 in a line so that the sum of each adjacent pair is a square number.
>
> (For example, 3 next to 6 is allowed because 9 is a square number, but 3 next to 5 is not allowed because 8 is not square.) Are some strategies better than others for solving this problem?

End of chapter reflection

You should know that...	You should be able to...	Such as...
In multiplication of positive and negative numbers: $+\times+=+$ $+\times-=-$ $-\times+=-$ $-\times-=+$ In division of positive and negative numbers: $+\div+=+$ $+\div-=-$ $-\div+=-$ $-\div-=+$	Multiply and divide positive and negative integers.	Find: **a)** $-8 \times (-7)$ -8×7 $8 \times (-7)$ **b)** $-64 \div (-8)$ $-64 \div 8$ $64 \div (-8)$ **c)** $-2 \times (-3) \times (-4)$ $-2 \times 3 \times (-4)$
-4^2 means square 4 and then subtract. $(-4)^2$ means square -4. -4^3 means cube 4 and then subtract. $(-4)^3$ means cube -4.	Square and cube positive and negative numbers, and understand the effect of brackets around the number.	Find the values of: **a)** -3^2 **b)** -5^3 **c)** $(-3)^2$ **d)** $(-5)^3$
A positive number has two square roots, one negative and one positive. A negative number has no square root.	Find both square roots of a positive number.	Write both square roots of 64.
A positive number has a positive cube root, and a negative number has a negative cube root.	Find cube roots of positive and negative numbers.	Write the values of: **a)** $\sqrt[3]{27}$ **b)** $\sqrt[3]{-27}$
A number with index 0 has the value 1.	Write the value of a number with index 0.	Find the values of: **a)** 7^0 **b)** 0.63^0 **c)** $(-20)^0$
You can use the index laws in some index calculations. The laws are: $n^a \times n^b = n^{(a+b)}$ $n^a \div n^b = n^{(a-b)}$	Use the index laws to simplify expressions involving powers.	Write as a single power of 7: **a)** $7^3 \times 7^6$ **b)** $7^9 \div 7^4$ **c)** $(7^4 \times 7^6) \div 7^5$

2 2D and 3D shapes

You will learn how to:

- Identify and describe the hierarchy of quadrilaterals.
- Understand that the number of sides of a regular polygon is equal to the number of lines of symmetry and the order of rotation.
- Understand and use Euler's formula to connect the number of vertices, faces and edges of 3D shapes.
- Understand π as the ratio between a circumference and a diameter. Know and use the formula for the circumference of a circle strands.

• •

Starting point

Do you remember…

- the terms 'parallel lines' and 'perpendicular lines'?

  For example, what is the name for lines that meet at right angles?

- the names of special triangles and quadrilaterals?

  For example, what types of quadrilaterals have four equal sides?

- how to find the lines of symmetry and order of rotational symmetry in polygons?

  For example, how many lines of symmetry in a rectangle?

  What is the order of rotational symmetry of a parallelogram?

- the names of common 3D shapes, and what vertices, edges and faces are?

  For example, name these 3D shapes and find how many vertices, edges and faces each one has.

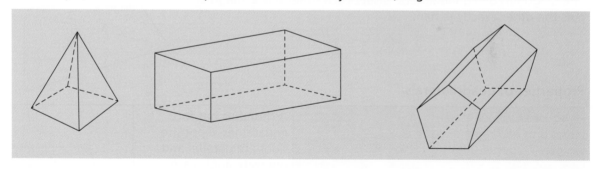

- the names of different parts of a circle?

  For example, what do you call the straight line segment that passes through the centre of a circle and whose endpoints lie on the circle? What is the special name given to the perimeter of a circle?

  What do you call any straight line segment that runs from the centre of a circle to any point on the edge of the circle?

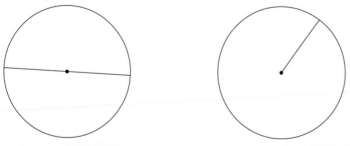

This will also be helpful when…

- you find the surface area and volume of cylinders.

Take any five points on the circumference of a circle and join them to make a five-sided polygon.
Two diagonals can be drawn from any vertex of the polygon.

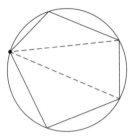

- How many diagonals can be drawn from one vertex in a seven-sided polygon?

- Find out how many diagonals could be drawn from **one** vertex in a polygon with 12 sides, without the need to draw it.

- How many diagonals **altogether** (from **all vertices**) can be drawn in a seven-sided polygon?

- How many diagonals **altogether** can be drawn in a 15-sided polygon, without the need to draw it?

2.1 Quadrilaterals

Key terms

A line segment drawn from one vertex of a quadrilateral to the opposite vertex is called a **diagonal**.

The diagonals of a quadrilateral **bisect** each other if one diagonal cuts the other exactly in half.

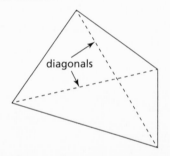

diagonals

Properties of quadrilaterals

A **square** has four equal sides and four right angles. Opposite sides are parallel. The diagonals of a square are equal in length, are perpendicular and bisect each other.		A **rhombus** has four equal sides. Opposite sides are parallel and opposite angles are equal. The diagonals of a rhombus are perpendicular and bisect each other.	
	A **rectangle** has four right angles. Opposite sides are parallel and equal in length. The diagonals of a rectangle are equal in length and bisect each other.		In a **parallelogram**, opposite sides are parallel and equal in length. Opposite angles are equal. The diagonals of a parallelogram bisect each other.

| A **trapezium** has one pair of parallel sides.

 In an **isosceles trapezium**, the two non-parallel sides are equal in length.

 The diagonals in an isosceles trapezium are equal in length. | 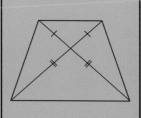 | A **kite** has two pairs of equal, adjacent sides.

 One pair of opposite angles are equal.

 The diagonals of a kite are perpendicular.

 One of the diagonals bisects the other. | 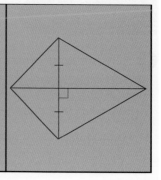 |

Hierarchy of quadrilaterals

You can see that:

All squares are special rectangles and rhombuses.

All rectangles are special parallelograms and trapeziums.

All rhombuses are special parallelograms and kites.

A parallelogram is a special trapezium.

> **Did you know?**
>
> If the opposite angles of a quadrilateral add up to 180°, then it is possible to draw a circle that passes through all four of the vertices.

Worked example 1

a) Abdul thinks of a quadrilateral. It has rotational symmetry and diagonals that are perpendicular. What types of quadrilateral could Abdul be thinking of?

b) Write down one property about the diagonals of a rectangle that parallelograms do not generally have.

| **a)** Square or rhombus | The types of quadrilateral with rotational symmetry are:

 Square   Rhombus
 Rectangle   Parallelogram

 Of these, the diagonals are perpendicular only in a square and a rhombus. | |
| **b)** The diagonals of a rectangle are the same length. | The diagonals of a rectangle bisect each other and are equal in length.

 One diagonal of a parallelogram is longer than the other. | |

1 Here are some properties relating to quadrilaterals.

Property A	Property B	Property C
Two pairs of equal sides.	At least one line of symmetry.	Diagonals that are equal in length.

Write down which of these properties, if any, the following quadrilaterals have.

a) kite   **b)** rectangle   **c)** trapezium   **d)** parallelogram   **e)** rhombus

2 Name the four types of quadrilateral where each diagonal bisects the other.

3 Write down a property about the diagonals of a square that is not shared by all rectangles.

4 Make a copy of this table.

	Diagonals are perpendicular	Diagonals are not perpendicular
Rotational symmetry		
No rotational symmetry		

Write the name of one type of quadrilateral in each cell of the table.

5 **a)** Decide whether each statement is true or false.

     (i) A square is also a rectangle.   **(ii)** A square is also a parallelogram.

     (iii) A rectangle is also a rhombus.   **(iv)** A kite also a parallelogram.

     b) Give a reason for each statement you have said is false.

6 Maria draws a quadrilateral with angles of 50°, 70°, 80° and 160°.

She says, 'My quadrilateral is a kite.'

Explain how you can tell that she is wrong.

7 The diagram shows two sides of a quadrilateral.

A teacher asks her class to complete the quadrilateral to make a shape with diagonals that are perpendicular.

Max says that it can't be done. Is he correct? Give a reason for your answer.

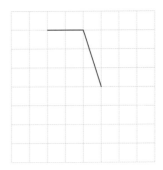

8 Explain why a rhombus is a special type of kite.

9 Helena says that all trapeziums are parallelograms.

a) Give a reason why Helena's statement is false.

b) Change Helena's statement so that it makes a true statement about trapeziums and parallelograms.

10 a) Rectangles have two lines of symmetry. Decide whether a shape must be a rectangle if it has this property. Give a reason for your answer.

b) Opposite sides in a rectangle are equal. Viktor says he can use this property on its own to test whether or not a shape is a rectangle. Comment on Viktor's test.

c) Find a single property that rhombuses have that can, on its own, be used as a test of whether or not a shape is a rhombus.

Thinking and working mathematically activity

Think about

Try to find a single property that a parallelogram has that can be used on its own as a test of whether or not a shape is a parallelogram. Do the same for other quadrilaterals.

Quadrilateral Parallelogram Isosceles trapezium
Rectangle Kite Rhombus Trapezium Square

- Which of the above shapes could also be a square?
- Which of the above shapes could also be a parallelogram?
- Which of the above shapes could also be a trapezium?
- Complete a copy of this diagram to show the hierarchy of the eight shapes.

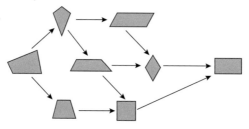

2.2 Polygon symmetry

 ### Thinking and working mathematically activity

- Describe the symmetries of an equilateral triangle by giving the number of lines of symmetry and order of rotational symmetry.
- Describe the symmetries of a square by giving the number of lines of symmetry and order of rotational symmetry.
- Describe the symmetries of a regular pentagon by giving the number of lines of symmetry and order of rotational symmetry.
- What is the link between the number of sides of a regular polygon and its symmetries?
- Describe the symmetries of a regular polygon with 50 sides.

Worked example 2

a) How many lines of symmetry does a regular heptagon have?

b) What is the order of rotation symmetry of a regular nonagon?

c) A regular hendecagon is an eleven-sided polygon. Describe the symmetries of a regular hendecagon.

a) seven	A regular heptagon has seven sides, the same number as the number of lines of symmetry.	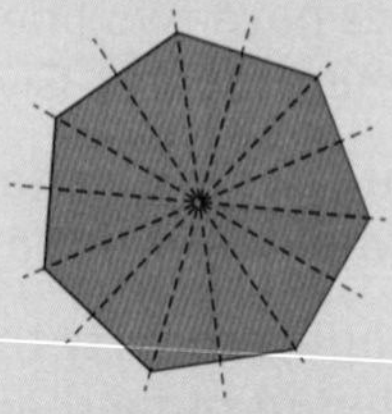
b) nine	A regular nonagon has nine sides, the same number as the order of rotational symmetry.	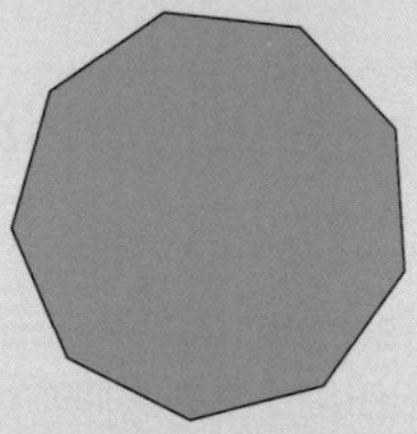
c) Eleven lines of symmetry and rotational symmetry of order eleven.	A regular hendecagon has eleven sides, the same number as the number of lines of symmetry and the order of rotational symmetry.	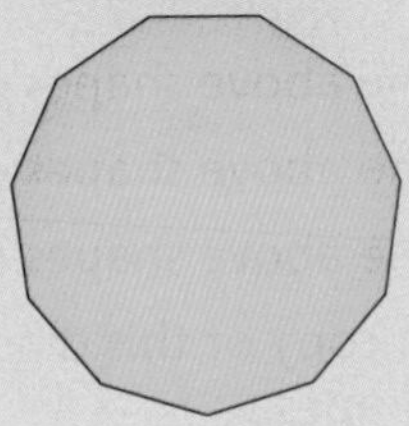

Exercise 2

1 Name a polygon with:

a) six lines of symmetry

b) rotational symmetry of order ten

2 State the number of lines of symmetry of:

a) a regular pentagon

b) a regular polygon with 14 sides

c) a regular polygon with rotational symmetry of order eight.

3 Explain why each statement is false.

a) A regular nonagon has 7 lines of symmetry.

b) A pentagon must have 5 lines of symmetry.

c) A shape with rotational symmetry of order 4 must be a square.

d) An octagon has rotational symmetry of order 8.

4 Decide whether the following statements are always true, sometimes true or never true?

Give a reason for each answer.

a) A triangle has 3 lines of symmetry.

b) A regular hexagon has 6 lines of symmetry.

c) A shape with 4 lines of symmetry also has rotational symmetry of order 4.

5 Amber says there is no shape with 13 lines of symmetry and rotational symmetry of order 13.

Is she correct? Explain your answer.

6 **a)** Draw a polygon with six lines of symmetry and rotational symmetry of order six.

b) How many other shapes can you draw with the same symmetries?

7 Ben says, 'I've drawn a shape with five lines of symmetry and rotational symmetry of order five.'

Mae says, 'You must have drawn a regular pentagon.'

Is Mae correct? Explain your answer.

2.3 3D shapes

Key terms

A **polyhedron** is a 3D shape with many flat faces. There are different types of **polyhedra**.

Some have flat bases rising to a single vertex, for example **pyramids** and **tetrahedrons**.

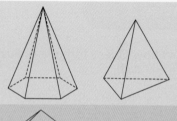

Some have a regular **cross-section**, for example **pentagonal prisms** and **triangular prisms**.

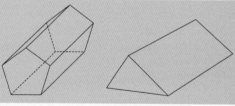

▼ Thinking and working mathematically activity

Here is a triangular prism. Write down the number of vertices (V), faces (F) and edges (E) for this prism.

- Think about some more polyhedra. For each of your polyhedra, write down the number of vertices (V), faces (F) and edges (E).

- Find a relationship between the number of vertices (V), faces (F) and edges (E). Your relationship should work for all of your polyhedra.

- Test to see if your relationship works on a different polyhedron.

- What have you discovered?

Your work in the activity will have lead you to find **Euler's formula** that connects the number of vertices, faces and edges of polyhedra.

Check your relationship is the same as Euler's formula which is **V + F − E = 2**

Worked example 3

a) You are told a polyhedron has 6 vertices and 8 faces. How many edges does it have?

b) How many vertices will there be in a 3D solid with 15 edges and 7 faces?

a) 12	Use Euler's formula: V + F − E = 2 V = 6 and F = 8 giving 6 + 8 − E = 2 14 − E = 2 So E = 12	
b) 10	Use Euler's formula: V + F − E = 2 E = 15 and F = 7 giving V + 7 − 15 = 2 V − 8 = 2 So V = 10	

Exercise 3

1 Show how Euler's formula is illustrated in each of these 3D shapes.

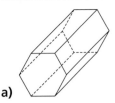
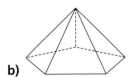

a)   b)   c)

2 Theo has a polyhedron with 8 vertices and 6 faces. Write down the number of edges it has.

3 Write down the number of vertices in a polyhedron with 8 edges and 4 faces.

4 The table shows some information about three polyhedra. Copy and complete the table.

	Number of vertices	Number of faces	Number of edges
Shape A	8		14
Shape B		8	10
Shape C	9	4	

5 Vitor said that an octahedron has 8 faces and 12 edges.

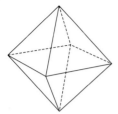

a) Is Vitor correct?

b) Write down the number of vertices an octahedron has.

6 Explain why it's impossible to make a polyhedron with 12 faces, 6 vertices and 10 edges.

7 Ben says he has a shape with 12 faces and 20 edges.

Dylan says it must have 10 vertices.

Is Dylan correct? Explain your answer.

2.4 Circumference of a circle

▼ Thinking and working mathematically activity

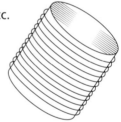

- Find as many cylinder shapes as you can: bottles, tins, cans, tubes, pipes etc.
- Measure the diameter of each item as accurately as you can.
- Now find the circumference of each one by the following method.

    Wrap a piece of string or cotton around the shape ten times.

    Make sure you have a start and a finish clearly marked.

    Unwind the string and measure the length from start to finish.

    This will be ten times the circumference. To find the circumference, divide this length by ten.

- Now copy and complete this table for your results.

Item	Circumference	Diameter	Circumference ÷ diameter

Complete the last column by dividing each circumference by its diameter – give the answers here correct to 3 decimal places.

- Write down what you notice about the numbers in the last column.
- What do you observe about the relationship between the circumference and diameter?

Circumference of a circle

The formulae to calculate the circumference C, of a circle are:

$$C = \pi d \text{ or } C = 2\pi r$$

where d is the diameter and r is the radius.

The value of π is 3.142 to 3 decimal places.

Did you know?

When we calculate the circumference of a circle, we use π (pi). There is a π button on a scientific calculator.

π has an infinite number of decimal places, so we will never know every single digit of π. The ancient Chinese were happy to use 3 as a value for π, and the Babylonians (about four thousand years ago) used the fraction $3\frac{1}{8}$.

By 1665, Sir Isaac Newton had calculated π to 16 decimal places.

Later, the use of computers meant that the number of known decimal places of π had increased to 2037 in 1949 and to greater than 1 million digits by 1973. By 2016, Peter Trueb had calculated π to almost 22.5 trillion digits!

Worked example 4

a) Find the circumference of a circle with radius 6 cm.

b) Find the diameter of a circle with circumference 50 cm.

Give your answers correct to 1 decimal place.

a) $C = 2\pi r$ $= 2 \times \pi \times 6$ $= 37.7$ cm (1 d.p.)	To work out circumference, use the formula $C = 2\pi r$ and substitute $r = 6$. Use a calculator to work out the circumference, giving the answer to 1 d.p.	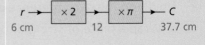
b) $C = \pi d$ $50 = \pi \times d$ So $d = 50 \div \pi$ $= 15.9$ cm (1 d.p.)	To work out circumference, use the formula $C = \pi d$ and substitute $C = 50$. The inverse operation to multiplication is division, so $d = C \div \pi$ Use a calculator to work out the diameter, giving the answer to 1 d.p.	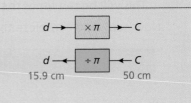

Exercise 4

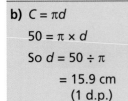

1 Find, correct to 1 decimal place, the circumference of a circle with diameter:

    a) 9 cm   **b)** 7.4 cm   **c)** 12.5 m   **d)** 53 mm

2 Find, correct to 1 decimal place, the circumference of a circle with radius:

    a) 5 cm   **b)** 4.2 cm   **c)** 6.1 m   **d)** 37 mm

3 Match each circle to the value of its circumference (rounded to 1 decimal place).

Could you have matched the correct answer to each of the pictures without using a calculator?

a)   b) 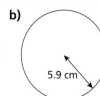 c) 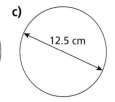 d)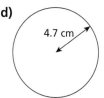

Circumferences:	
A 37.1 cm	B 29.5 cm
C 27.0 cm	D 39.3 cm

4 Copy and complete the table.

Radius	Diameter	Circumference of circle (rounded to 1 decimal place)
6.8 cm		
	11.6 cm	
2.1 m		
	134 mm	

5 For each statement below, say whether it is always true, sometimes true or never true. Give a reason for each answer.

a) The circumference of a circle is greater than its diameter.

b) The circumference of a circle is greater than 3 cm.

c) The radius of a circle is greater than the circumference.

6 A rope is wound 18 times around a large drum of diameter 45 cm. Find the length of the rope.

7 The minute hand of a clock is 6 cm long.

Andrew said the tip of the hand will move 38 cm in one hour.

Is Andrew correct? Give a reason for your answer.

8 Find, correct to 1 decimal place, the diameter of a circle with circumference:

a) 29 cm   b) 78 cm   c) 12 m   d) 128 mm

9 Find, correct to 1 decimal place, the radius of a circle with circumference:

a) 38 cm   b) 97 cm   c) 21 m   d) 217 m

> **Think about**
>
> There many fractional approximations to π.
>
> Probably the most famous is $\frac{22}{7}$
>
> Others include $\frac{47}{15}$, $\frac{69}{22}$ and $\frac{113}{36}$
>
> Investigate online to see if you can find a fractional approximation to π that is correct to 4 decimal places.

10 A bicycle wheel has diameter 70 cm.

    a) Find how far the bicycle travels if the wheel makes 20 revolutions.

    b) Find the number of complete revolutions the wheel makes when the bicycle travels 1 km.

11 A cycle track is circular. The inner radius is 60 m and the track is 5 m wide.

Kat cycles on the outside edge of the track.
Eve cycles on the inside edge of the track.

Find how much further Kat cycles than Eve in making one circuit of the track.

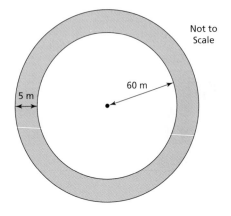

Not to Scale

60 m

5 m

Consolidation exercise

1 Write down the name of the quadrilateral that is being described in each part. There may be more than one answer for each.

    a) Opposite sides are equal in length. The diagonals bisect each other and are equal in length.

    b) Opposite sides are parallel. Every side is the same length.

    c) The diagonals intersect at right angles with the shorter diagonal bisected.

    d) Opposite sides are parallel and equal in length.

    e) Each side is the same length. Each angle is 90°.

2 Ollie describes a type of quadrilateral:

'The diagonals bisect each other. It has less than two lines of symmetry.'

    a) Write down the name of two possible quadrilaterals it could be.

    b) Write down another property that would help find what the shape actually is.

3 Match each description to the correct quadrilateral.

Diagonals are equal in length and perpendicular.	Rectangle
Diagonals are perpendicular but not equal in length.	Square
Diagonals are equal in length but not perpendicular.	Parallelogram
Diagonals are not equal in length and are not perpendicular.	Kite

4 Decide if these statements about quadrilaterals are always true, sometimes true or never true.
If the statement is sometimes true, draw a diagram to explain your answer.

Statement 1: Trapeziums have a line of symmetry.

Statement 2: Kites have one right angle.

Statement 3: Trapeziums have a pair of parallel sides.

Statement 4: The diagonals divide a square into equilateral triangles.

5 Name the following polygons.

a) It has rotational symmetry of order 5, and 5 lines of symmetry.

b) It has 7 lines of symmetry and rotational symmetry of order 7.

6 Rio draws a regular polygon. He says it has 6 lines of symmetry and rotational symmetry of order 3.

Can Rio be correct? Give a reason for your answer.

7 A polyhedron has 8 faces and 9 edges. Find how many vertices it has.

8 Sophia says she has made a polyhedron with 4 faces, 3 vertices and 6 edges.

Could she be correct? Give a reason for your answer.

9 Dylan says he has a polyhedron with 8 faces, 7 vertices and 10 edges.

Dylan has made a mistake.

Two of his values are correct.

State the possible correct number of faces, vertices and edges.

10 Suha has a cake shop. She is baking cakes with a diameter of 22 cm. She is going to wrap some ribbon around the outside of the cakes. How many cakes can Suha decorate with 5 m of ribbon?

11 Letsaby is making a piece of jewellery.

The design is made from five circles made of wire. The four small circles are all the same size.

The large circle has a diameter of 5 cm.

Letsaby has 1 m of wire. She thinks she can make three pieces of jewellery

with this design. Is Letsaby correct? Explain how you know.

End of chapter reflection

You should know that...	You should be able to...	Such as...
There is a hierarchy of quadrilaterals.	Identify quadrilaterals from their properties.	Name the quadrilateral that has two pairs of equal sides and perpendicular diagonals where the longer one bisects the shorter.
The number of sides of a regular polygon is equal to the number of lines of symmetry and the order of rotational symmetry.	Identify a regular polygon by knowing one symmetry property.	Name the regular polygon with six lines of symmetry.
Euler's formula ($V + F - E = 2$) connects the number of vertices, faces and edges of 3D shapes.	Find the number of faces, vertices or edges in a 3D shape when given the values of the other two.	Find the number of edges are there in a 3D shape that has 5 faces and 7 vertices.
π is the ratio between a circumference and a diameter. The formula for the circumference of a circle is $C = \pi d$.	Calculate the circumference of a circle given either the radius or the diameter. Calculate the diameter or radius of a circle given the circumference.	Find the circumference of a circle with radius 6 cm. Find the diameter of a circle with circumference 65 cm.

3 Collecting data

You will learn how to:

- Select, trial and justify data collection and sampling methods to investigate predictions for a set of related statistical questions, considering what data to collect (categorical, discrete and continuous data).
- Understand the advantages and disadvantages of different sampling methods.

Starting point

Do you remember…

- that data can be collected using different methods?

  For example, using observation, interviews and questionnaires to collect data.

- how to take a random sample from a population?

  For example, taking a random sample using random numbers or drawing names from a hat.

- that sample size can affect the accuracy of your results?

  For example, a sample of size 50 will typically give more accurate results than one of size 10.

This will also be helpful when…

- you learn about bias in data collection and sampling methods
- you carry out your own statistics projects.

3.0 Getting started

How many students in your class walk to school?

Use this information to estimate the number of students in your school who walk to school.

How accurate do you think your estimate is? What factors could affect the accuracy of your estimate?

Decide whether your results will be similar for other schools in the country.

3.1 Data collection methods

Key terms

Primary data is information collected by the person doing the investigation. The data could be collected from an experiment or a survey.

Secondary data is information collected by someone else. It could be taken from the internet, books, academic journals, newspapers, etc.

Worked example 1

A company wants to build a shopping centre in a town. They want to find out whether the people who live in the area would use the shopping centre and which shops they would like to be in it.

The researcher is trying to decide whether to conduct face-to-face interviews or send out a questionnaire in the post.

Comment on the advantages and disadvantages of each of the approaches.

Advantages of interviews The researcher can clarify the question if people are unclear. **Disadvantages of interviews** People may not be as honest during a face-to-face interview. It may be time consuming to interview lots of people. It will cost more to employ people to conduct the interviews.	The researcher can think about: • Time – How long will it take to collect the data? • Cost – How expensive will it be to collect the data? • Honesty – Will people be more or less likely to tell the truth using each method? • Number of responses – Will people answer the questions?
Advantages of a postal questionnaire The researcher can send the questionnaire to a lot of people. People can take their time when completing the questionnaire. **Disadvantages of a postal questionnaire** People may not complete it or forget to return it. You cannot explain the questions to the person answering the questionnaire.	**Think about** Are there any other advantages or disadvantages that you can think of?

![arrow banner] **Exercise 1**

1 **Vocabulary question** Write down whether each person is using a primary or secondary data source.

a) Lucy records the number of pages in books on her bookcase.

b) Max measures the length of bananas.

c) Natalia finds the masses of the planets orbiting the sun in a textbook.

d) Ori asks a sample of people how much they get paid.

e) Pietra downloads the results for a swimming race from the internet.

▼ **Thinking and working mathematically activity**

• Can you think of a research question that can be answered using either primary or secondary data?

• Can you think of a research question that you can only answer using primary data? Explain why this can only be answered using primary data.

• Can you think of a research question that you can only answer using secondary data? Explain why this can only be answered using secondary data.

2 Andrei and Cathy want to find out how many people would be interested in joining a film club at school.

Andrei says, 'We should ask a sample of 20 people from our classes.'

Cathy says, 'We should ask a sample of 100 people in the school canteen.'

Comment on which approach would be better and give a reason for your answer.

3 Valentina wants to find out the views of people in her town about plans to re-open an old cinema. She decides to send out 1000 questionnaires to people in the town.

Write down whether each statement about the questionnaires is true or false.

a) It will be less time consuming than doing face-to-face interviews.

b) Valentina will definitely get all of the questionnaires back.

c) The answers will be more accurate than if she does face-to-face interviews because she can explain the questions if there is any confusion.

4 Filip wants to investigate the following research question:

Do snakes live longer in the wild or in captivity?

a) Which two of the following would it be most useful to collect data about?

A	B	C	D
Age of the snake	Colour of the snake	Where the snake lives	Length of the snake

b) Which two of the following methods would be best for Filip to collect his data?

Method 1: Buy a pet snake and see how long it lives.

Method 2: Use the internet to find the lifespan of different species of snake.

Method 3: Conduct interviews with friends and family to ask their opinion.

Method 4: Contact several experienced snake handlers and conduct telephone interviews.

5 Zlakto wants to find out where students in his school go on holiday. Do you think he should: (a) collect data from the internet, (b) send out a questionnaire to 100 randomly selected students, or (c) ask every student in his school? Give a reason for your choice.

6 Molly wants to compare the number of men and women living in her country. Explain why it will be more convenient for Molly to use a secondary source of data.

7 Kalandra wants to investigate how the maximum daily temperature in her village varies throughout the year.

a) Write down one specific question that Kalandra could investigate.

b) Kalandra finds some temperature data for a nearby city on the internet.
Would this be primary or secondary data?

c) Why might Kalandra get more useful data if she recorded the temperature each day herself?

8 Greg is the manager of a swimming pool. He wants to find out people's views about the changing facilities at the pool. He decides to do this using a questionnaire.

He thinks of two possible questions that he could ask to get people's views:

Question A
What do you think about the changing facilities at the pool? Tick a box.
Very poor ☐   Poor ☐   Satisfactory ☐   Good ☐   Very good ☐   Don't know ☐

Question B
What do you think about the changing facilities at the pool?
....
....
....

Write down the advantages and disadvantages of the two questions.

9 Henri wants to interview a sample of 200 people in a city about their use of bikes.
He is considering these two methods.

Method A	Method B
Interview people on the telephone.	Interview people face-to-face.

a) Give one reason why Henri may want to use Method A.

b) Give one reason why he may want to use Method B.

10 Kira wants to investigate the amount of time people spent at work last week.

a) Suggest two different methods that she could use to collect relevant data.

b) Which of your two methods would you recommend to Kira? Give a reason for your choice.

> **Did you know?**
>
> The first time a survey was used to predict the result of an election was in the United States in 1824.

Worked example 2

200 lorry drivers and 50 office staff work for a transportation company.

The company is thinking of changing the number of days of holiday staff can have.

The company wants to find out the views of its workers.

They decide to use one of the following methods to select a sample of 30 workers:

      Method 1: Randomly choose 24 of the lorry drivers and 6 of the office staff.

      Method 2: Choose 30 staff eating in the canteen.

Give one advantage and one disadvantage of each method.

Method 1	Remember that a sample should be representative of the population it is chosen from.
Advantage: This method has four times as many lorry drivers in the sample as office staff. This means the sample will have a similar structure to the population.	It is often a good idea to choose people in the sample using a random selection method. This helps to ensure that the sample will be representative.
Disadvantage: It might take time to track down all the selected workers. The lorry drivers may be out on the road.	
Method 2	Choosing people who are easy to pick makes doing the sampling straightforward. But the people selected may not be representative of the population.
Advantage: This method will be quick and easy as it will be convenient to speak to people when they are in the canteen.	
Disadvantage: There may not be many lorry drivers in the canteen as they could be driving. The people asked might be mainly office staff.	

Exercise 2

1 James wants to investigate the times taken by men and women to run a marathon.

A website gives the times for 30 000 men and 20 000 women to run the marathon.

a) Suggest one reason why James may want to use all the available data.

b) Suggest one reason why he might instead want to take a sample of men and a sample of women.

2 There are 500 trees in a park. Macik wants to investigate their heights by taking a sample of 40 trees.

He can use either of these sampling methods:

      Method 1: Measure the heights of the first 40 trees he sees.

      Method 2: Measure the heights of a random sample of 40 trees in the park.

Which method should he use? Give a reason for your choice.

3 Moira is a supermarket manager. She wants to ask customers about the freshness of the vegetables on sale in her supermarket.

She decides to carry out face-to-face interviews with customers.

She decides to select customers using one of these two methods.

> **Method 1:** Ask 10 customers each day for one week, interviewing at different times of the day.

> **Method 2:** Ask 80 customers using the supermarket on a Thursday evening.

Which data collection method is best? Give a reason for your answer.

4 A school has 400 boys and 100 girls. The headteacher wants to find out the views of a representative sample of students about school clubs.

She decides to write a questionnaire. She plans to choose people to answer her questionnaire using one of these three methods.

Method A	Method B	Method C
Ask for 100 volunteers to stay behind after assembly to complete the questionnaire.	Ask 80 boys and 20 girls chosen at random to complete the questionnaire.	Ask the 100 students who come first alphabetically on the school register.

Which method is best? Give reasons for your answer.

5 Gina works in a biscuit factory. She wants to check that biscuits made by a machine have the correct mass.

She can use either of the following sampling methods.

> **Method 1:** Record the mass of every 100th biscuit made by the machine.

> **Method 2:** Choose one packet of biscuits and find the mass of every biscuit in that packet.

Which method should she choose? Give a reason for your answer.

6 A football club wants to find out the views of fans about the price of tickets. It considers asking 40 people attending one match.

a) Suggest a better way for the club to get the information it wants.

b) Explain why your method should give the club more reliable information.

▼ Thinking and working mathematically activity

Technology question A theatre has 225 seats. Every seat in the theatre is sold on one particular day. The ages of the people in the theatre are shown in the seating plan on the next page.

1	2	3	4	5	6	7	8	9	10	11	12	13	14	15
45	9	10	10	9	9	9	24	10	10	10	9	9	9	43
16	17	18	19	20	21	22	23	24	25	26	27	28	29	30
39	10	9	10	10	9	54	9	9	10	9	9	10	9	31
31	32	33	34	35	36	37	38	39	40	41	42	43	44	45
42	9	9	10	9	10	10	45	9	10	9	10	10	9	39
46	47	48	49	50	51	52	53	54	55	56	57	58	59	60
59	9	10	9	9	25	10	10	9	9	41	9	10	10	30
61	62	63	64	65	66	67	68	69	70	71	72	73	74	75
9	9	10	36	24	43	12	12	13	12	12	13	12	12	27
76	77	78	79	80	81	82	83	84	85	86	87	88	89	90
26	12	13	12	13	12	13	13	12	12	12	13	12	13	47
91	92	93	94	95	96	97	98	99	100	101	102	103	104	105
12	13	12	13	31	12	13	14	48	12	13	13	12	12	13
106	107	108	109	110	111	112	113	114	115	116	117	118	119	120
45	37	46	48	39	62	63	61	36	10	12	35	29	8	7
121	122	123	124	125	126	127	128	129	130	131	132	133	134	135
76	74	77	48	50	33	12	36	56	57	55	51	63	31	28
136	137	138	139	140	141	142	143	144	145	146	147	148	149	150
15	16	42	77	80	34	36	69	64	92	89	53	54	31	33
151	152	153	154	155	156	157	158	159	160	161	162	163	164	165
12	14	47	44	59	53	38	11	14	8	43	75	73	44	42
166	167	168	169	170	171	172	173	174	175	176	177	178	179	180
11	11	45	77	78	75	55	56	29	47	17	15	14	52	13
181	182	183	184	185	186	187	188	189	190	191	192	193	194	195
37	44	39	35	40	13	11	38	60	55	33	35	21	20	20
196	197	198	199	200	201	202	203	204	205	206	207	208	209	210
17	15	43	74	70	66	69	35	34	41	43	28	29	21	21
211	212	213	214	215	216	217	218	219	220	221	222	223	224	225
65	67	72	69	75	64	75	81	83	56	64	63	69	41	43

Key:

Seat number
Age

The manager wants to find an estimate of the mean age of the people at the theatre.

He considers the following methods to obtain a sample of people.

Method 1: Choose everyone sitting in one row.

Method 2: Choose people sitting on every 15th seat, starting with seat 1.

Method 3: Choose 5 people at random from those people in the theatre.

Method 4: Choose 15 people at random from those people in the theatre.

- Which method do you think is best? Why do think this?
- Collect the sample of ages the manager gets from each of these sample methods.
- Find the mean age for each sample.

- Think about the advantages and disadvantages of the different methods.
- The mean age for everyone in the theatre is actually 31.2 years.
  Which sampling method gave a mean value closest to this?
- Compare your results with those from other people in your class.

Consolidation exercise

1 Five people are using sets of data.

Nina collects data from a wildlife magazine that shows the number of lions living in different countries.

Evie measures the height of her sunflower every morning.

Will looks at newspaper reports showing the wealth of the world's 10 richest people in 2019 and 2020.

Hester asks 100 people how they plan to vote in an election.

Naga uses the internet to find the number of goals scored in 20 matches last season.

Which people are using primary data and which are using secondary data?

2 Liam designs an experiment to investigate whether young people can solve a word puzzle quicker than older people.

a) He records the time that people take to solve the puzzle. What other information does Liam need to record?

b) Is Liam collecting primary or secondary data? Give a reason for your answer.

3 A local football club wants to ask the views of some of its 12 000 season ticket holders.

Do you think they should conduct face-to-face interviews with each season ticket holder, or send out questionnaires to a sample of 1000 season ticket holders?

Give a reason for your choice.

4 Angelica is a dentist. She wants to collect some information from her patients about what they think about the service she offers.

She will give the questionnaire to a sample of patients using one of these methods.

Method A	Method B
Send an email to all her patients with a link to a questionnaire to complete online.	Ask all the patients that she sees one week to complete a questionnaire on their way out.

Which method would you recommend Angelica uses? Give a reason for your choice.

5 A tennis club has 300 members. There are 100 children and the rest are adults.

The club secretary wants to choose a sample of 15 people to ask their opinion about where to hold the end of year party.

a) She first thinks about asking 15 members playing tennis one Monday morning.
Explain why this would not give her a representative sample.

b) She then decides to ask 10 adults and 5 children chosen at random.
Explain why this is a better way to choose her sample.

6 Finn wants to ask a sample of people in a stadium about where they live and how they got there.

He is considering these sampling methods.

Method 1: Selecting the first 100 people to leave the stadium

Method 2: Selecting 100 people queuing for food

Method 3: Choosing 100 seats at random and asking the people sitting at those seats

Which sampling method would you recommend Finn to use?

Give a reason for your choice.

> **Discuss**
>
> What are the disadvantages of the methods you did not choose?

End of chapter reflection

You should know that...	You should be able to...	Such as...
Data can be collected from primary or secondary sources.	Tell the difference between primary and secondary data.	A doctor measures the blood pressure of 20 patients. Does the doctor collect primary or secondary data?
A range of factors need to be considered when choosing a method for collecting data, including things like: • Cost • Time • Convenience • Quality of the results obtained	Decide which method of data collection is appropriate. Compare two or more data collection methods.	Give the advantages and disadvantages of: • face-to-face interviews • postal questionnaires
A good sample should be representative of the population it is taken from. The choice of sampling method will depend on things such as the time available and how reliable the results need to be.	Decide which method of sampling is appropriate. Compare two or more sampling methods.	Which of these methods for getting a sample of students from a school is likely to give better results? • Choosing everyone in one class • Choosing a sample at random from everyone in the school

4 Factors and rational numbers

You will learn how to:

- Understand factors, multiples, prime factors, highest common factors and lowest common multiples.
- Understand the hierarchy of natural numbers, integers and rational numbers.

Starting point

Do you remember…

- how to find lowest common multiples and highest common factors for numbers below 100?

  For example, find the LCM and HCF of 45 and 60.

- the difference between composite numbers and prime numbers?

  For example, state which of these are prime numbers: 21, 23, 25, 27, 29, 31, 33, 35, 37, 39

This will also be helpful when…

- you learn about irrational numbers.

4.0 Getting started

Writing messages in code is called cryptography. It is particularly important for sending private messages, such as bank account details, between computers.

Prime numbers can be used in cryptography. For example, the sender's computer finds the product of two large prime numbers, and then rewrites the message using a code based on the product.

The receiver's computer has information that tells it how to decode the message. But if anyone else reads the coded message, they cannot decode it unless they can factorise the product. This is very difficult to do, even for a computer.

Each red number below is the product of two prime numbers.

6   21   34   39   77   91   209   299

Using the prime numbers in the grid, see if you can find which two of them multiply to give each number.

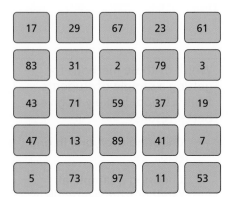

17	29	67	23	61
83	31	2	79	3
43	71	59	37	19
47	13	89	41	7
5	73	97	11	53

> ### Did you know?
>
> If a very large number is the product of two prime numbers, it is extremely difficult to find the prime numbers. A 240-digit number was 'cracked' into its two prime factors in 2019. It took two years, using the equivalent of 450 advanced computers.

Key terms
....

A **prime number** is a number with exactly two factors: 1, and the number itself. This means that 1 is **not** a prime number. The first seven prime numbers are: 2, 3, 5, 7, 11, 13, 17.

A **composite number** is an integer that has more than two factors. Examples are: 4, 6, 8, 9, 12, 14, 15.

The **prime factors** of a positive integer are the prime numbers that multiply to make the integer. Every integer has a unique set of prime factors. For example, the prime factors of 60 are 2 × 2 × 3 × 5. These can be written using **index notation** as $2^2 \times 3 \times 5$.

Worked example 1

Write 120 as a product of its prime factors.

Find the prime factors of 120.	Draw a factor tree by following the steps below. Write 120 and draw lines to two integers that multiply to make 120, for example, 2 and 60. Repeat for 60 (but not for 2, which is prime). Repeat for 30, and then 15. At the end, all branches should end in prime numbers. These are the prime factors of 120.	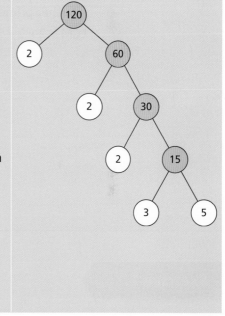
120 = 2 × 2 × 2 × 3 × 5 120 = $2^3 \times 3 \times 5$	Write the prime factors in increasing order. You can also write them using index notation.	

Worked example 2

Use prime factorisation to find the highest common factor (HCF) and lowest common multiple (LCM) of 20 and 88.

20 = 2 × 2 × 5	Find the prime factors of each number.	Factor tree for 20:

88 = 2 × 2 × 2 × 11

Factor tree for 88:

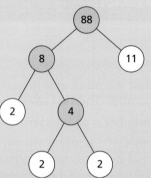

The HCF of 20 and 88 is 4. The LCM of 20 and 88 is 440.	The HCF is the product of the common (shared) prime factors: HCF = 2 × 2 = 4 The LCM is the product of the common prime factors (2 and 2), and all other prime factors (2, 5 and 11): LCM = 2 × 2 × 2 × 5 × 11 = 440	You can use a Venn diagram to find the HCF and LCM. Write the prime factors of 20 in one circle and the prime factors of 88 in the other. Write common prime factors in the intersection (overlapping area). The HCF is the product of the numbers in the intersection: 2 × 2 = 4 The LCM is the product of all the numbers in the diagram: 2 × 2 × 2 × 5 × 11 = 440

Exercise 1 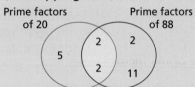 1–5

1 Which of these are prime numbers?

31   33   43   47   51   59   77   79

2 Which of these statements are true and which are false?
Explain your answers.

a) 6 is a factor of the number 2 × 3 × 3

b) 10 is a factor of the number $2 × 3^2$

c) 12 is a factor of the number 2 × 2 × 3

d) 12 is a factor of the number $2^2 × 4$

e) 15 is a factor of the number 2 × 3 × 5 × 7 × 13

f) 40 is a factor of the number 2 × 2 × 5

3 Find the number with prime factor products:

a) $2^2 \times 7$ b) $2^2 \times 3 \times 5$ c) 5×7^2

d) $2^3 \times 3^2$ e) $2^3 \times 5^3 \times 7$ f) $2^2 \times 3 \times 11$

Tip

Remember that you can change the order of numbers in a multiplication.

4 Write each number as the product of its prime factors. Use index form where appropriate.

a) 50 b) 72 c) 92 d) 105

e) 108 f) 120 g) 125 h) 148

i) 160 j) 171 k) 177 l) 198

5 Below are the prime factor products of two numbers. List all the factors of each number.

a) $2 \times 3 \times 5$ b) $2 \times 5 \times 5 \times 7$

6 Find the HCF and LCM of each pair of numbers. The prime factor products of each number are provided to help you.

a) $50 = 2 \times 5 \times 5$
   $70 = 2 \times 5 \times 7$

b) $66 = 2 \times 3 \times 11$
   $110 = 2 \times 5 \times 11$

c) $90 = 2 \times 3 \times 3 \times 5$
   $117 = 3 \times 3 \times 13$

d) $121 = 11 \times 11$
   $132 = 2 \times 2 \times 3 \times 11$

e) $65 = 5 \times 13$
   $130 = 2 \times 5 \times 13$

f) $104 = 2 \times 2 \times 2 \times 13$
   $135 = 3 \times 3 \times 3 \times 5$

7 Use prime factorisation to find the HCF and LCM of:

a) 8 and 20 b) 32 and 24 c) 50 and 60

d) 54 and 45 e) 70 and 55 f) 72 and 80

Tip

In question 7, try sketching a Venn diagram of the prime factors.

8 Find two factors of 108 such that their sum is double their difference.

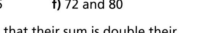

9 Two numbers have HCF = 5 and LCM = 60.

a) Ellie thinks that one of the numbers is 15. What is the other number?

b) Jake thinks that there is another pair of numbers with HCF 5 and LCM 60. What could those numbers be?

10 An integer n is divisible by 6.

Write down two prime factors of n.

Think about

The number 2 is the only even prime number. Can you explain why?

Discuss

A number x is made by multiplying three prime numbers, 11, 17 and 23. List all the factors of x.

Thinking and working mathematically activity

Find the prime factors of some square numbers and some numbers that are not squares.

What do you notice about the prime factors of square numbers? Write a method for using prime factors to check to see if a number is a square number. Can you explain why your method works?

Investigate the prime factors of cube numbers.

Use prime factorisation to find which of the numbers below are squares or cubes.
Find the square roots of the square numbers, and the cube roots of the cube numbers.

900   1160   1296   1728   2025   2304   4096   5832

Which number is both a square and a cube? How do its prime factors show this?

4.2 Rational numbers

Key terms

The **natural numbers** are the positive integers: 1, 2, 3, 4, 5, ...

A **rational number** is a number that can be written as a fraction. For example, 0.17 is a rational number as it can be written as $\frac{17}{100}$.

Did you know?

The number π (pi) cannot be written as a fraction. It is an example of an 'irrational number'.

Worked example 3

Write down the numbers from the list:

-12 $\frac{1}{2}$ 0.75 π 100

that are

    a) rational numbers

    b) integers

    c) natural numbers.

a) -12, $\frac{1}{2}$, 0.75, 100	A rational number is any number that can be written as a fraction. These numbers can be written as fractions: $-12 = -\frac{12}{1}$ $\frac{1}{2}$ is already written as a fraction $0.75 = \frac{75}{100}$ $100 = \frac{100}{1}$ π cannot be written exactly as a fraction, so it is not rational.
b) -12, 100	An integer is a whole number. It can be negative or positive.
c) 100	A natural number is a positive integer.

1 Below is a list of numbers.

-88 -17.2 $-\dfrac{50}{7}$ $-\pi$ $\dfrac{1}{3}$ 1 7.43 68

Write down which numbers are:

a) rational numbers   **b)** integers   **c)** natural numbers

2 In part **a)** and part **b)**, use the clues to find the number n.

a) n is an integer greater than -1. It is not a natural number.

b) n is a natural number between $\dfrac{5}{2}$ and 3.85

3 **a)** Write a natural number that lies between -7 and 2

b) Write an integer that lies between -7.73 and -6.41

4 For each pair of rational numbers, find another rational number that lies between them.

a) $\dfrac{1}{4}$ and $\dfrac{7}{8}$ **b)** 3.1 and 4.2   **c)** 0.78 and 0.79   **d)** 0.999 and 1

5 Is each statement always, sometimes or never true?

If a statement is always true or never true, explain why.

If a statement is sometimes true, write one example where it is true and one example where it is not true.

a) The product of two integers is a natural number.

b) The quotient of two integers is an integer.

c) The quotient of two integers is a rational number.

d) The product of two fractions is a rational number.

e) The quotient of two fractions is a rational number.

> **Tip**
>
> The quotient of two numbers is the result when you divide the first number by the second number.

6 **Vocabulary question** Copy the sentences below and fill in the blanks using the list of words in the box.

positive   negative   rational   integers   natural

Numbers that can be written as fractions are called .... numbers. These include positive and .... whole numbers, which are called .... whole numbers are called .... numbers.

▼ Thinking and working mathematically activity

A rational number is any number that can be written as a fraction. Investigate this idea, exploring what types of number you can write as fractions.

Write a short guide to recognising rational numbers.

Draw a Venn diagram showing rational numbers, natural numbers and integers. Explain how you drew it.

How many different regions does your diagram have? Write two numbers in each region.

1 **a)** Find the prime factors of:

 i) 24   **ii)** 32

 Use your answers to solve the problems below.

 b) Ben has two pieces of ribbon. One is 32 cm long. The other is 24 cm long. He cuts them into smaller pieces of equal length, with no ribbon left over. What is the greatest length of each piece in cm?

 c) Lily needs to buy cups and plates for a party. There are 24 cups in a pack. There are 32 plates in a pack. She needs exactly the same number of cups as plates. What is the smallest number of each pack she can buy?

2 Find four two-digit numbers with exactly:

 a) Two prime factors

$$\boxed{}\boxed{} = \boxed{} \times \boxed{}$$

 b) Three prime factors

$$\boxed{}\boxed{} = \boxed{} \times \boxed{} \times \boxed{}$$

 c) Four prime factors

$$\boxed{}\boxed{} = \boxed{} \times \boxed{} \times \boxed{} \times \boxed{}$$

 d) Five prime factors

$$\boxed{}\boxed{} = \boxed{} \times \boxed{} \times \boxed{} \times \boxed{} \times \boxed{}$$

3 In the diagrams below, the number in the middle square is the HCF of the numbers opposite each other. Use the numbers 8, 18, 24, 30, 40, 54, 60 to complete the diagrams.

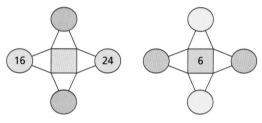

> **Tip**
>
> Think about how you use prime factors to find the HCF and LCM.

4 Is the following statement true or false? Explain your answer.

 The product of two numbers is equal to the product of their HCF and LCM.

5 Are these statements true or false?

 a) 7 is a rational number.

 b) −7 is a rational number.

 c) 7 is a natural number.

 d) −7 is a natural number.

 e) 7 is an integer.

 f) −7 is an integer.

6 Write down:

 a) a rational number x, greater than −1 but less than 0

 b) a natural number n, greater than −4 but less than 4

 c) an integer m, greater than −3 but less than 3

End of chapter reflection

You should know that...	You should be able to...	Such as...
Every positive integer can be written as a product of prime numbers. These are called the prime factors of the number.	Find the prime factors of a number and write them using index notation.	Find the prime factors of: **a)** 48　　**b)** 120 **c)** 68　　**d)** 108
Prime factors can be used to find the highest common factor and lowest common multiple of two or more integers.	Use prime factors to find the HCF and LCM of two numbers.	Use prime factors to find the HCF and LCM of 48 and 120.
A rational number is a number that can be written as a fraction.	Recognise rational numbers.	Which of these numbers are rational? -1.5, $\frac{4}{5}$, 2, π, 17
An integer is a positive or negative whole number. A natural number is a positive whole number.	Recognise integers and natural numbers.	Which of these numbers are integers, and which are natural numbers? -18, -7, -1, 6, 13

Expressions

You will learn how to:

- Understand that letters have different meanings in expressions, formulae and equations.
- Understand that the laws of arithmetic and order of operations apply to algebraic terms and expressions (four operations, squares and cubes).
- Understand how to manipulate algebraic expressions including:
  - applying the distributive law with a single term (squares and cubes)
  - identifying the highest common factor to factorise.
- Understand that a situation can be represented either in words or as an algebraic expression, and move between the two representations (linear with integer or fractional coefficients).

Starting point

Do you remember...

- that you can use letters to represent unknown numbers, variables or constants?

  For example, 'a number plus three' can be written as $n + 3$

- that an algebraic expression is made up of terms?

  For example, the expression $2x + 3y + 1$ is made up of three terms.

- that the coefficient of a variable is the number it is multiplied by?

  For example, the term $7t$ means '$7 \times t$', where 7 is the coefficient of the variable t.

- that you can substitute numbers into expressions?

  For example, you can work out $5s + t$ when $s = 4$ and $t = 3$

  $5s + t = 5 \times 4 + 3$

  $ = 20 + 3 = 23$

This will also be helpful when...

- you substitute values into expressions and formulae
- you solve linear equations
- you simplify and construct more complex expressions
- you factorise expressions by writing them with brackets.

5.0 Getting started

The expression in each box is the sum of the expressions in the two boxes below it.

- Complete the number pyramid to find the expression in the top box.
- What other expressions can you get in the top box by rearranging the four expressions in the bottom row?
- Can you find four new expressions to put into the bottom row that make the top expression equal to $21n + 7$?

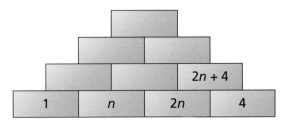

Key terms

An **expression** connects numbers and **variables** with mathematical operations (such as +, −, × and ÷). For example, $\frac{x}{2} + 2y - 5$. The value of this expression can be found for any given values of x and y.

An **equation** contains an equals sign. It is a statement that shows that the two expressions on either side of the equals sign are equal. For example, $7x - 1 = x + 5$. In this equation, there is only one possible value of x. x is an **"unknown"** but the equation can be solved to find what x is ($x = 1$).

A **formula** is a type of equation that expresses the relationship between certain variable quantities. For example, the area of a rectangle = length × width or $A = lw$. The value of A can be found for any given values of l and w.

$y = 3x - 2$ is also a **formula** that describes the relationship between the variables x and y. It is also sometimes known as a **linear function.** A linear function is a function whose graph is a straight line. y is called the **subject of the formula** as the formula shows how to work out a value of y from a given value of x.

Did you know?

We use equations and formulae to describe real world situations. They help us understand what is going on and to make predictions. For example, by considering the forces acting on a bridge, the engineers and architects decide on the structure they will use to support the bridge. The choice of material depends on a wide range of factors.
These factors include the load to be carried (including the weight of vehicles), the span length of the bridge, the construction and maintenance costs, the visual impact and so on.

Worked example 1

Is each of these an expression, an equation or a formula?

a) $5n + 2 = 17$ **b)** $4(2a + b)$

c) $V = abc$, where V = volume, a = length, b = width and c = height

a) equation	$5n + 2 = 17$ contains an equals sign and the variable n represents a particular value that could be found ($n = 3$).	$\boxed{n}\ \boxed{n}\ \boxed{n}\ \boxed{n}\ \boxed{n}\ \boxed{2}$ $\boxed{\qquad 17 \qquad}$
b) expression	$4(2a + b)$ is an expression containing brackets. If the brackets are expanded, the expression is equivalent to $8a + 4b$.	$\boxed{a}\,\boxed{a}\,\boxed{b}\ \boxed{a}\,\boxed{a}\,\boxed{b}\ \boxed{a}\,\boxed{a}\,\boxed{b}\ \boxed{a}\,\boxed{a}\,\boxed{b}$ $2a + b \quad 2a + b \quad 2a + b \quad 2a + b$
c) formula	$V = abc$ is a formula. It expresses the relationship between the volume of a cuboid and the cuboid's measurements.	

1 Is each of these an expression, an equation or a formula?

a) $17 - 3m$ b) Charge = 25 × number of hours   c) $5n = 35$

d) $y = 2x + 1$ e) $\frac{a}{4} + 2b - 3$ f) $\frac{x}{2} = 11$

2 In which of these does n represent a particular fixed value?

Expression $30 - n$	Formula $y = 2n$
Equation $2n - 3 = 7$	Formula The cost \$C of buying n books is $C = 12n$

3 Which of these are formulae?

a) $u = 11 - t$ b) $y + 2x - 3$ c) $h + 5 = 2h$ d) $y = 3x + 7$

> **Think about**
>
> Is y the subject of the formula in $2x + 10 = y$?

4 In which of these is y the subject of the formula?

$y = x - 2$ $y + 2x = 5$ $y + x = 9$ $y = 3 + 2h$

5 Technology question

a) Use the internet to find a formula for:

   i) the average speed of an object, using the distance it travels and the time it takes

   ii) the area of a trapezium

   iii) the force generated by a pressure acting over a specified surface area.

b) Find some other interesting formulae and write them down. Explain what they help you find.

6 Nafia says that $17 = 24 - n$ is a formula.
Explain why Nafia is incorrect.

7 Write down four formulae that have 'x' as their subject.

> **Tip**
>
> How many possible values of n can you find?

▼ Thinking and working mathematically activity

Decide if these statements are true or false. Give a reason to explain your answers.

- An equation always contains an addition (+) or a subtraction (−) sign.
- An equation always contains an equals sign (=).
- If there is an equals sign (=) in an algebraic relationship, then it must be an equation.

Make up some more statements about expressions, formulae and equations. Some of your statements should be true and some should be false. Swap with a partner and see if they can tell which are the true statements.

5.2 Algebraic operations and substitution

Key terms

Algebraic operations must follow the same order as those in arithmetic.

For example,

in $2a^3$ the power is calculated first

in $4(d - 3)$ the brackets are expanded first

in $\dfrac{3a + 5}{2}$ the numerator is considered first.

The order in which operations must be applied is summarised as **BIDMAS**:

	Brackets first
then	**I**ndices
then	**D**ivision and **M**ultiplication
and finally,	**A**ddition and **S**ubtraction

Worked example 2

Find the value of:

    a) $5x^2 + 3$, when $x = -2$

    b) $\dfrac{2y + 1}{3}$, when $y = 1$

    c) $2(p - q^3)$, when $p = 4$, $q = 2$

a) $5x^2 + 3$ $= 5(-2)^2 + 3$ $= 5(4) + 3$ $= 20 + 3$ $= 23$	Remember the order of operations. First find the value of x^2 Next, find the value of $5x^2$ by multiplying the value of x^2 by 5 Finally, add 3 to get the value of $5x^2 + 3$
b) $\dfrac{2y + 1}{3}$ $= \dfrac{2 \times 1 + 1}{3}$ $= \dfrac{2 + 1}{3}$ $= \dfrac{3}{3}$ $= 1$	First find the value of $2y$, by multiplying 2 by y Add 1 to the value of $2y$ Divide the value of the numerator by 3
c) $2(p - q^3)$ $= 2(4 - 2^3)$ $= 2(4 - 8)$ $= 2 \times (-4)$ $= -8$	Start with the brackets – find the value of q^3 first. Next, find the value inside the brackets by subtracting the value of q^3 from the value of p. Multiply the value inside the brackets by 2.

1 When $x = -6$, find the value of:

   a) $x - 10$ **b)** $15 - x$ **c)** $4x$ **d)** $3x + 5$ **e)** $5 - 4x$

2 When $x = -2$, find the value of:

   a) $2x$ **b)** $3x^2$ **c)** $4 - x^2$ **d)** $5 + 2x^2$ **e)** x^3

> **Tip**
>
> $x^3 = x \times x \times x$

3 When $p = 2$ and $q = -3$, find the value of:

   a) $p + q$ **b)** $p - q$ **c)** $2p - q$ **d)** $20 + 2q$

   e) $5 - 3q$ **f)** q^2 **g)** $2p^2$ **h)** q^3

   i) $p^2 - q^2$ **j)** $p^2 + q^2$ **k)** $p^3 - q^3$ **l)** $3p^3$

> **Tip**
>
> Ensure you use the correct order of operations (BIDMAS).

4 When $a = 6$, $b = -2$ and $c = -5$, find the value of:

   a) $a + 2b$ **b)** $b + c$ **c)** $b - c$ **d)** $a + b + c$ **e)** $ab + c$ **f)** abc

   g) $2b^2$ **h)** $2b^2 + c$ **i)** $a^2 - b^2$ **j)** $b^3 + 5$ **k)** $b^3 - a$ **l)** $b^2 - 2c^2$

5 In each part, write down the part of the calculation that should be performed first.

   a) $3x + 5$ multiplication or addition

   b) $\dfrac{x}{4} - 2$ division or subtraction

   c) $2(x - 4)$ multiplication or brackets

   d) $4a^3$ multiplication or power

   e) $(x - 2)^2$ brackets or power

6 Calculate the following when $x = 5$, $y = 4$ and $z = 6$

   a) $x(y + z)$ **b)** $xy + z$ **c)** $(x + y)^2$ **d)** $z^2 + x^2$

   e) $\dfrac{yz}{x}$ **f)** $\dfrac{y^3 - 2x}{z}$ **g)** $3x^2 - 2yz$ **h)** $7(4z - x^2)$

7 Write these expressions in ascending order (smallest to largest).

   Use the values $x = 3$, $y = -2$ and $z = -1$.

   xy^2 x^2y $2yz^2$ $3x^2 + y$ $2y^2 - z$

▼ Thinking and working mathematically activity

   Decide if these statements are always true, sometimes true or never true. Explain your answers.

- $2x$ is more than x
- x^2 is more than x
- x^3 is more than x

8 Jake finds the value of $3x^2 - 2y$ when $x = 1$ and $y = 3$.

These are his calculations.

$3 \times 1^2 - 2 \times 3$

$3 \times 1 = 3$

$3^2 - 2 \times 3$

$9 - 2 \times 3$

$7 \times 3 = 21$

His answer is wrong.

Show Jake how he could have worked out the correct answer to this calculation.

5.3 Expanding and factorising

Key terms

Brackets are **expanded** (or removed) by multiplying the term outside the brackets by the expression on the inside. The reverse of expanding brackets is **factorising**.

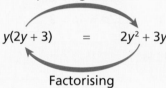

Expanding brackets

$$y(2y + 3) \quad = \quad 2y^2 + 3y$$

Factorising

Tip

Expand means 'multiply out'. You multiply each term inside the brackets by the term outside. To factorise an expression look for the highest common factor which divides all the terms.

Worked example 3

Expand the brackets:

a) $n(n - 3)$　　**b)** $5p(2p + 3q)$　　**c)** $3p^2(p - 2q)$

a) $n(n - 3)$ $= n^2 - 3n$	Multiply both the terms inside the brackets by the term on the outside.		n	$-$	3
		n	n^2	$-$	$3n$

b) $5p(2p + 3q)$ $= 10p^2 + 15pq$	Multiply both terms inside the brackets by $5p$. $5p \times 2p = 5 \times 2 \times p \times p = 10p^2$ $5p \times 3q = 5 \times 3 \times p \times q = 15pq$		$2p$	$+$	$3q$
		$5p$	$10p^2$	$+$	$15pq$

c) $3p^2(p - 2q)$ $= 3p^3 - 6p^2q$	Multiply both terms inside the brackets by $3p^2$ $3p^2 \times p = 3 \times p \times p \times p = 3p^3$ $3p^2 \times 2q = 3 \times 2 \times p^2 \times q = 6p^2q$		p	$-$	$2q$
		$3p^2$	$3p^3$	$-$	$6p^2q$

Worked example 4

a) Factorise $18n - 6$

b) Factorise $a^2b + 7ab$

c) Factorise $8pq - 12p$

a) $18n - 6$ $\quad = 6 \times 3n - 6 \times 1$ $\quad = 6(3n - 1)$	Find the HCF of $18n$ and 6. It is 6. Rewrite each expression as a product of 6. Put 6 outside the brackets and find what needs to go on the inside to make it multiply out correctly.	
b) $a^2b + 7ab$ $\quad = ab \times a + ab \times 7$ $\quad = ab(a + 7)$	Find the HCF of a^2b and $7ab$. It is ab. Rewrite each expression as a product of ab. Put ab outside the brackets and find what needs to go inside.	$a^2b = ab \times a$ $7ab = ab \times 7$
c) $8pq - 12p$ $\quad = 4p \times 2q - 4p \times 3$ $\quad = 4p(2q - 3)$	The HCF of $8pq$ and $12p$ is $4p$. Rewrite each expression as a product of $4p$. Take $4p$ outside the brackets as a factor and find what should go inside.	$8pq = 4p \times 2q$ $12p = 4p \times 3$

Exercise 3

1 Expand the brackets:

a) $7(2a + 3b)$ b) $y(y + 9)$ c) $n(n - 6)$ d) $c(4c + 3d)$

e) $3g(g - 4)$ f) $5r(2r + 11)$ g) $2tu(6t - 3v)$ h) $6p(3p - 2q + 4)$

2 a) Find three equal pairs of expressions from those below:

$6r(2r + 3)$ $8r^2 + 18r$ $3r(4r + 3)$ $6r + 9$

$2r(4r + 9)$ $12r^2 + 9r$ $8r + 18$ $12r^2 + 18r$

b) Write an equivalent expression for each of the two remaining expressions.

3 Expand and then simplify each expression.

a) $5(2t + 5) - 3t$ b) $6 + 3(2x - 7) + x$ c) $9(4m - n) + 3n$ d) $8(3e + f) - 4e - 11f$

4 Find the odd one out in the expressions below. Give reasons for your answer.

$4n(2n + 1)$ $6n^2 + 4n$ $2n(4n + 2)$ $8n^2 + 4n$

5 Explain the mistake that has been made in each expansion.

a) $4(2n + 5) = 6n + 20$ **b)** $t(t + 3) = 2t + 3t$ **c)** $3r(2r - 5) = 6r^2 - 5$

6 Expand and simplify:

a) $3(4q - 7) + 2(9 - q)$ **b)** $2t(t - 5) - t^2 + 6t$

c) $4w(3w + x) + w - 5wx - w^2$ **d)** $3y(3y - 2) + 2y(7 - y)$

7 Match each pair of expressions with their highest common factor.

a^2b and $8a^2$	a
ab and $4a$	4
$2a^2b$ and $3ab$	$4a$
$8a$ and $12b$	a^2
$12a$ and $16ab$	ab

Thinking and working mathematically activity

* Find two expressions which have an HCF equal to $4p$. Can you find another pair?
* Can you find three expressions which have an HCF equal to $4p$?
* How many pairs of expressions can you find which have an HCF equal to $3pq$?

---

8 Find the missing terms:

a) $3t \times ..... = 15t$ **b)** $9y \times ..... = 9y^2$ **c)** $6r \times ..... = 12r^2$ **d)** $..... \times ab = 3a^2b$ **e)** $3pq \times ..... = 9p^2q$

9 Complete these factorisations:

a) $12y - 32 = ... (3y - 8)$ **b)** $5p + 15q = ... (p + ...)$ **c)** $18c - 6 = ... (...... - 1)$

d) $t^2 + t = ... (...... + 1)$ **e)** $2a^2 + ab = ... (2a + ...)$ **f)** $5f^2 + 7f = ... (...... + 7)$

10 Factorise these expressions:

a) $16e - 24$ **b)** $9y + 18z$ **c)** $21 - 14gh$ **d)** $12p + 24q - 6$

e) $h^2 - h$ **f)** $4xy + 3x$ **g)** $5m^2 + 6m$ **h)** $8a - ab - a^2$

11 Factorise these expressions:

a) $6r^2 - 10r$ **b)** $8pq + 12pr$ **c)** $9k^2 - 3k$ **d)** $15wx - 10wy^2$

e) $ab^2 - 2ab$ **f)** $6a^3 - 9a^2$ **g)** $4t^2 + 2t - 12tu$ **h)** $12n^2m - 8nm^3$

12 Use the expressions from the box to complete these statements:

.... + ... = ... (....... + ...)

.... − .... = 6(....... − ....)

2g	3g	4g	15g
	24g	6g²	
	5	3	18

13 Sonia has tried to factorise three expressions. Explain what mistake she has made in each one.

a) $4g + 6 = 4(g + 2)$

b) $2a^2 − 3a = a(2^2 − 3)$

c) $18h^2 + 45h = 3h(6h + 15)$

14 **a)** This is a rectangle.

2x

2x + y

Write down which of these are possible expressions for the rectangle's area.

$2x(2x + y)$ $8x + 2y$ $4x^2 + 2xy$ $4x^2 + 2x + y$

b) This is a different rectangle.

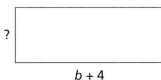

?

b + 4

An expression for the area of this rectangle is $2b^2 + 8b$.

Write down which of these expressions must be the rectangle's width.

$b + 4$ $2b$ $4b$ $b + 2$

Worked example 5

A shop sells two different packets of biscuits.

Small   **Large**

Contains n biscuits
Cost x dollars

Contains 10 more biscuits than a small packet
Cost y dollars

Sara buys three small packets and two large packets.

Find an expression for:

a) the total cost

b) the total number of biscuits.

a) $(3x + 2y)$ dollars	The cost of three small packets is $3 \times x = 3x$ dollars. The cost of two large packets is $2 \times y = 2y$ dollars.	x | x | x | y | y $3x$ | $2y$ $3x + 2y$
b) $3n + 2(n + 10)$ 　$= 3n + 2n + 20$ 　$= 5n + 20$	A large packet contains 10 more biscuits than a small packet. So a large packet contains $n + 10$ biscuits. Two large packets will contain $2(n + 10) = 2n + 20$ biscuits.	n | n | n | $n + 10$ | $n + 10$ $3n$ | $2n + 20$ $5n + 20$

1. Form a simplified expression for the perimeter of each shape.
   All lengths are measured in centimetres.

   a)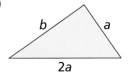

   b)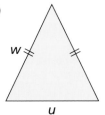

   c)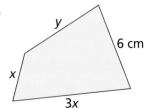

   d)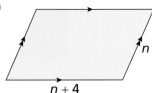

2. A plant pot costs \$x. A plant costs \$w.

   Find an expression for:

   a) the difference between the cost of a plant pot and the cost of a plant

   b) the total cost of a plant pot and 8 plants.

3. The image shows the costs of tickets at a cinema.

   Write down an expression for:

   a) the cost of 4 adult tickets

   b) the total cost of 2 child tickets and 1 adult ticket

   c) the total cost of 3 child tickets, 4 student tickets and 2 adult tickets

   d) the amount of change from \$50 if 3 adult tickets are bought.

4. A pizza costs \$p plus the cost of the toppings. Each topping costs \$t.

   a) Find the cost of a pizza with 2 toppings.

   b) Freddie orders a pizza with 3 toppings. Gina orders a pizza with 1 topping.
      Find the total cost of the two pizzas.

5. Martha has 2 litres of water. She fills 4 glasses. Each glass holds g litres of water.

   a) Write an expression for the amount of water she has left.

   b) Find the value of the expression from part a) when $g = 0.3$.

   c) What does your answer represent?

6. A baker has y grams of flour. A loaf of bread uses 400 grams of flour.

   a) How much flour does the baker have left after making n loaves?

   b) Find the amount of flour that is left when $y = 8200$ and $n = 15$.

7 A shop sells three types of badge.

A **red** badge has mass r grams.

The mass of a **blue** badge is 4 grams more than the mass of a red badge.

The mass of a **green** badge is three times the mass of a blue badge.

Find, in terms of r, an expression for:

a) the mass of a green badge

b) the total mass of 5 blue badges

c) the total mass of 6 red badges and 2 green badges.

8 Liam has m pens. Malik has twice as many pens as Liam. Noor has 3 more pens than Malik.

Omar has twice as many pens as Noor. Find an expression in terms of m for:

a) the number of pens than Noor has

b) the number of pens that Omar has

c) the difference between the number of pens Omar and Liam have

d) the total number of pens the four children have.

9 A packet contains n balloons. Nat buys 3 packets of balloons and shares them equally between 10 people.

a) How many balloons does each person receive?

b) Find the number of balloons they each receive when $n = 40$.

10 Cakes are sold in three different packs.

Small packs contain c cakes.

Medium packs contain 4 more cakes than a small pack.

Large packs contain three times as many cakes as a small pack.

a) Safee buys $6c + 12$ cakes. Which packs could she have bought?

b) Thimba buys $9c + 4$ cakes. He buys 5 packs in total. Which packs does he buy?

11 Shami has n socks.

Tom has six fewer socks than Shami.

Ursula has four more socks than Shami.

Victor has twice as many socks as Tom.

a) Form an expression for the total number of socks that the four people have.

b) Form an expression for the mean number of socks they have.

c) Find the mean number of socks if Shami has 10 socks.

12 Alexei has two bags of sweets. One bag contains n sweets. The second bag contains $n + 20$ sweets. One third of his sweets are strawberry flavoured. Find an expression for the total number of strawberry sweets he has.

13 All the classes in a school have either 30 children or 29 children. There are x classes with 30 children and y classes with 29 children.

   a) Find an expression for the total number of children in the school.

   b) There are g girls in the school. Find an expression for the number of boys.

   c) There are 15 times more children in the school than teachers. Find an expression for the number of teachers.

14 Small loaves of bread are made using p grams of flour. Large loaves of bread are made using $2p$ grams of flour. A baker makes $2n$ small loaves and $3n$ large loaves every day. Find an expression for the amount of flour used in 5 days.

15 Amol, Bea and Cadi are saving money for a holiday.

   They start with $\$b$ in a bank account.

   They each save $\$n$ each week.

   The holiday costs $2000.

   What does each of these expressions represent in this context?

   a) $2000 - b$ **b)** $3n$ **c)** $2000 - b - 3nw$

16 Jamila has $\$y$. She spends half of her money on a mobile phone. She then buys k maps costing $\$m$ each. Find an expression for the amount of money she has left.

17 Monty has n boxes. Eight of his boxes each have mass a kg. His other boxes each have mass b kg.

   a) Find an expression for the total mass of all of his boxes.

   b) Find an expression for the mean mass of his boxes.

▼ Thinking and working mathematically activity

a) Tim thinks of a number n. He adds 4. He multiplies the answer by 3. He then divides the answer by 5. Tim says his answer is $\dfrac{n + 12}{5}$.

   What mistakes has Tim made? What should the answer be?

b) Jade says that the area of a rectangle with side lengths $3a$ and $\frac{1}{2}b$ is represented by $3a \times \frac{1}{2}b = 3\frac{1}{2}ab$. What mistake has Jade made?

c) Work in pairs. Write two formulae that have a mistake in them. Swap the formulae with your partner. Find the mistakes and correct them.

---

1 Erin is e years old.

Fabio is $e + 5$ years old.

Gabriela is $3e + 15$ years old.

Henri is $3e + 13$ years old.

Complete these sentences by writing in either a number or an expression.

a) Gabriela is …….. times as old as Fabio.

b) Henri is ….. years younger than Gabriela.

2 Copy and complete the table to show equivalent expressions.

	Expression with brackets	Expression without brackets
a)	$q(q - 3)$	$q^2 \ldots \ldots$
b)	$2x(3x + 7)$	
c)	$4y(2y + 5z - 7)$	
d)	$\ldots \ldots .(3d + 10)$	$12d^2 + \ldots \ldots$

3 Match each expression with its simplified form.

$5x - 2y + x - 7y$	$9y - 2x$
$2x + 11y - 4x - 2y$	$9y - 6x$
$6y - 2x - 3y - 4x$	$6x - 9y$
$x + 8y - 5x + y - 2x$	$3y - 6x$

4 Insert the terms into the statements beneath to make them true. Each expression should be used exactly once. Can you complete the expressions in a different way?

$$4x \quad 5x \quad 7x \quad 2y \quad 6y \quad 7y$$

$6x - \ldots\ldots\ldots + \ldots\ldots\ldots + \ldots\ldots\ldots = 11x + 5y$

$\ldots\ldots\ldots - 5y - \ldots\ldots\ldots + \ldots\ldots\ldots = y - 3x$

5 How many factorised expressions can you find that are equivalent to $24x^2y + 18y^2$?

6 Choose the correct complete factorisation of each expression:

a) $12t - 6$

b) $5a^2 + 4a$

c) $4n^2 - 6mn$

$3(4t - 3)$	$6(2t - 0)$	$6(2t - 1)$
$a(5^2 + 4)$	$a(5a + 4)$	$a(5a + 3a)$
$n(4n - 6m)$	$2n(2n - 3m)$	$4n(n - 2m)$

7 Jan has n small glasses and m large glasses which she fills with water. A small glass holds 200 ml of water. She uses c litres of water altogether.

Explain what each of these expressions represents in this context.

a) $1000c$ **b)** $1000c - 200n$ **c)** $\dfrac{1000c - 200n}{m}$

End of chapter reflection

You should know that...	You should be able to...	Such as...
A variable is a letter that represents a number that can take different values. A formula is a type of equation that gives the relationship between variable quantities.	Recognise expressions, equations, and formulae.	A shop uses this rule to calculate the charge \$$C$ of delivering n items: $C = 15 + 3n$ Is this an expression, an equation or a formula?
Substitution is when you replace a variable with a known value.	Substitute positive and negative numbers into expressions and formulae.	If $A = 5p^2 - q$, calculate the value of A when $p = 3$ and $q = -9$
The order for performing operations in algebra and arithmetic is given by BIDMAS.	Recognise the order in which operations are performed.	In the expression $3(g - 2)$, which operation is done to g first: $- 2$ or $\times 3$?
You can expand a set of brackets by multiplying all the terms inside the brackets by the expression on the outside.	Expand a set of brackets.	Expand: $5f(2f - 3)$
Factorising involves putting brackets into an expression.	Factorise an expression by taking out a single-term common factor.	Factorise $3y^2 + 11y$
You can construct expressions using letters to represent unknown quantities.	Form an expression to represent a situation.	A tank contains t litres of water. Every hour n litres of water is taken out. Write an expression for the amount of water in the tank after 5 hours.

Angles

You will learn how to:

- Recognise and describe the properties of angles on parallel and intersecting lines, using geometric vocabulary such as alternate, corresponding and vertically opposite.
- Derive and use the fact that the exterior angle of a triangle is equal to the sum of the two interior opposite angles.
- Understand and use bearings as a measure of direction.

Starting point

Do you remember…

- the conventions for labelling points, sides and angles of shapes?

  For example, in a triangle *ABC*, what are the names for the two sides that form angle *ABC*?

- how to find a missing angle in a quadrilateral and in a triangle?

  For example, if three angles in a quadrilateral are 100°, 75° and 30°, what is the fourth angle?

- the sum of the angles at a point on a straight line is 180°?
  For example, find the missing angle in this diagram.

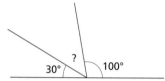

- the sum of the angles around a point is 360°?
  For example, what is the angle labelled *x* in this diagram?

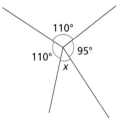

- the properties of angles on parallel lines and transversals, perpendicular lines and intersecting lines?
  For example, which of these angles are equal to each other?

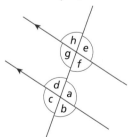

- the four points of the compass?
  For example, North is one compass point. Name the other three.

Naz is investigating the angles formed by the diagonals of quadrilaterals.

He starts with a square and notices that the diagonals form four right angles.

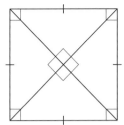

- Which other quadrilaterals have diagonals that meet at right angles?
- Can you find a quadrilateral that has diagonals that do not meet at right angles? What angles do the diagonals form?
- What do you notice about the sum of the angles produced at the point where the diagonal of a quadrilateral intersect? Is this always true? How do you know?
- Naz notices that drawing on the diagonals of a square produces four small triangles that meet at the centre.

What is the sum of all twelve of the angles in these four triangles?

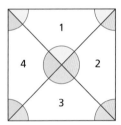

- Use the angle sum from the four triangles and the angle sum at the point where the diagonals intersect to show that the angles in a square must add up to 360°.
- Use a similar approach to show that the angles in a parallelogram must add up to 360°.

6.1 Special angles

Key terms

When two lines intersect, they form a pair of **vertically opposite angles** that are equal to each other.

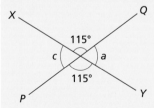

The angles marked 115° are vertically opposite and equal.
The angles a and c are also vertically opposite and equal.

When a transversal crosses two (or more) parallel lines it forms sets of equal angles.

Corresponding angles appear on the same side of the transversal:

Alternate angles appear on opposite sides of the transversal:

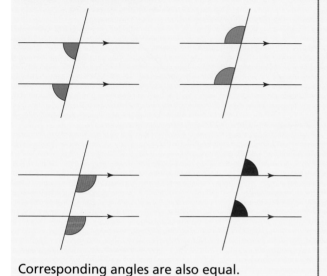

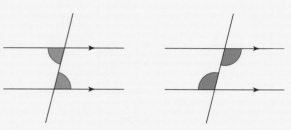

Alternate angles are **equal**.

Corresponding angles are also equal.

Worked example 1

a) What name is given to the two shaded angles?

b) Find the sizes of angles a, b and c

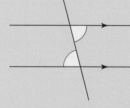

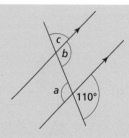

a) The angles are alternate angles.	The shaded angles form a reverse Z-shape. They are on the opposite sides of the transversal, so they are alternate angles.	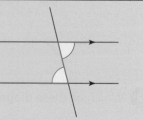

b) $a = 110°$	The angle marked a is vertically opposite to the angle 110°. Vertically opposite angles are equal, so $a = 110°$	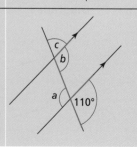
$b = 110°$	The angle marked b is corresponding to the 110° angle (the angles form an F-shape)	
$c = 70°$	The angles marked b and c form a straight line so they add up to 180° So $c = 180 - 110 = 70°$	

1 Write down whether the marked angles are vertically opposite angles, alternate angles or corresponding angles.

a)

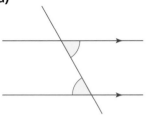

b)

c)

d)

e)

f)

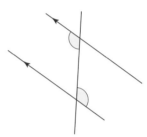

g)

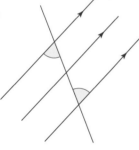

h)

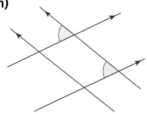

i)

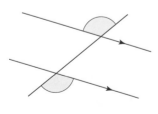

2 Which of these diagrams show corresponding angles?

Diagram 1

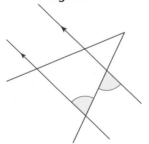

Diagram 2

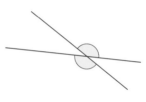

Diagram 3

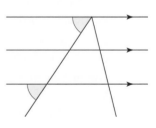

3 State which of these diagrams show alternate angles.

Diagram 1

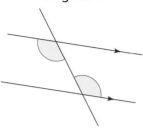

Diagram 2

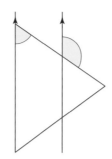

Diagram 3

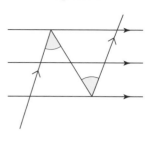

4 Find the size of each angle marked with a letter. For each answer, say whether you are using vertically opposite angles, corresponding angles, alternate angles or angles on a line.

a)

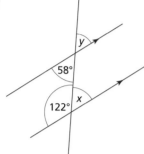

b)

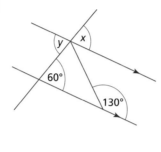

c)

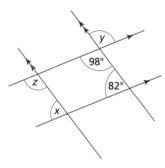

d)

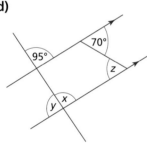

e)

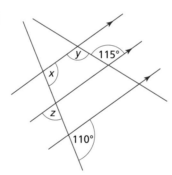

5 Here is a diagram involving parallel lines.

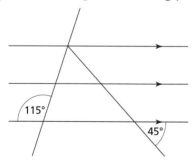

Copy the diagram and mark on all the angles equal to 115° and all the angles equal to 45°.

6 The diagram is formed from two sets of parallel lines.

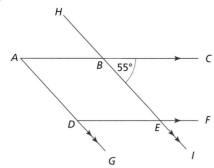

Say whether each statement is true or false.

a) Angles *CAG* and *FDG* are corresponding angles.

b) Angle *HBC* = angle *BED*

c) Angles *BED* and *EDG* are alternate angles.

d) Angle *BED* = 55°

e) Angles *ADF* and *ABE* are corresponding angles.

7 Find the size of each angle marked with a letter. For each answer, say whether you are using vertically opposite angles, corresponding angles or alternate angles.

a)

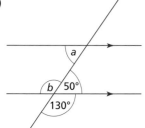

b)

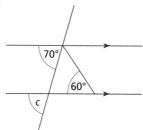

c)

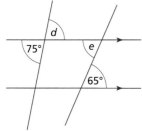

d)

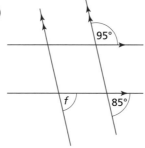

8 The diagram shows angles marked *a* to *k*.

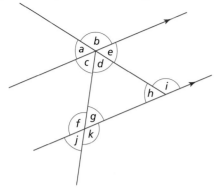

Write down the letter of the angle that is:

    a) vertically opposite to *f*
    b) alternate to *g*
    c) corresponding to *b*
    d) corresponding to *a*
    e) alternate to *h*
    f) corresponding to *j*
    g) vertically opposite to *e*

9 In the diagram below, are the lines *AB* and *CD* parallel?
Explain your answer.

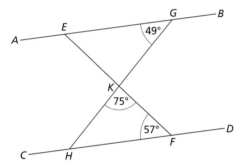

10 The diagram shows two right-angled triangles, *ABC* and *BCD*. Copy the diagram and fill in the sizes of all the marked angles.

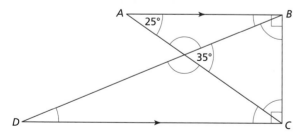

11 Vocabulary question Copy and complete the text using words from the box.

> corresponding   alternate   vertically opposite
> equal   parallel   transversal

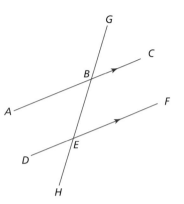

The lines *AC* and *DF* are _____ lines. They are crossed
by the line *GH* which is called a _____ .
Angles *ABH* and *GEF* are _____ and known as _____ angles.
Angles *ABH* and *CBG* are known as _____ angles.
An example of a pair of _____ angles are angles *CBH* and *FEH*.

▼ Thinking and working mathematically activity

In this diagram, angles *CBE* and *FEB* are known as co-interior angles. What is the relationship
between them?

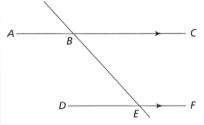

How would you explain to someone else what co-interior angles are?

6.2 Angles of a triangle

Key terms

The angles inside a shape are called **interior** angles.

An **exterior** angle of a shape is the angle between one side of the shape and the line created if the
next side is extended outwards.

For example:

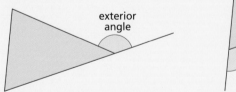

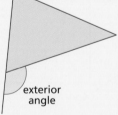

> **Think about**
>
> Is there only one position for the exterior angle at each vertex? If there is more than one
> possibility, where else could the exterior angle be?

Worked example 2

Complete this proof to show that the exterior angle of a triangle is equal to the sum of the two interior opposite angles.

Here is a triangle, with interior angles of size a and b.

The final interior angle of the triangle is an angle of size c.

a) Write a formula for c in terms of a and b.

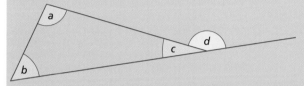

An exterior angle of the triangle of size d is shown in the diagram.

b) Write a formula connecting angles c and d.

c) Use the two formulae to show that $d = a + b$.

a) The final interior angle of the triangle, c, has size $180 - a - b$ because the angles in a triangle sum to $180°$.	We do not know the sizes of angles a and b but we find angle c by subtracting these from $180°$.
b) Angles c and d lie on a straight line, and the sum of angles on a straight line is $180°$. So the exterior angle, $d = 180 - c$	
c) Using the two expressions above: $d = 180 - (180 - a - b)$ $d = 180 - 180 + a + b$ $d = a + b$	Substituting for c Expanding the brackets Simplifying

> **Tip**
>
> You should be able to explain each step of your working out using a geometric reason or a rule.

Worked example 3

Find the size of the angles marked with letters in each diagram, stating the geometric reason for each step of your findings.

a)

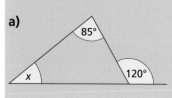

b)

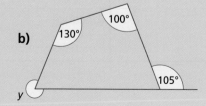

a) $x + 85 = 120$ because the exterior angle of a triangle is equal to the sum of the two interior opposite angles. $\qquad x = 120 - 85$ $\qquad = 35°$	Use the fact that the exterior angle of a triangle is equal to the sum of the two interior opposite angles.	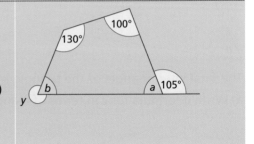
b) $a = 75°$, because angles on a line add up to $180°$ $b = 55°$, angles in a quadrilateral add up to $360°$ $y = 305°$, because angles around a point add up to $360°$	Label the two missing angles inside the quadrilateral a and b. $a = 180 - 105 = 75°$ $b = 360 - (75 + 100 + 130)$ $\qquad = 55°$ $y = 360 - 55$ $\qquad = 305°$	

Exercise 2

1 Complete this proof to show that the sum of the interior angles in a triangle is 180°.

Here is a triangle with internal angles a, b and c and a line drawn parallel to the base through the top vertex. Two further angles, d and e, have been labelled.

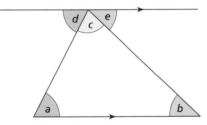

1 $a = d$	(because ....)
2 $b = e$	(because ....)
3 So the sum of the angles in the triangle $\quad = a + b + c$ $\quad = \ldots + \ldots + c$	(substituting for a and b using steps **1** and **2** above)
4 $= 180°$	(because ....)

2 Find the size of the angles marked with letters in each diagram, stating the geometric reason for each step of your findings.

a)

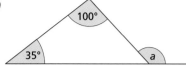

b)

3 Find the size of the angles marked with letters in each diagram, stating the geometric reason for each step of your findings.

a)

b)

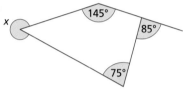

c)

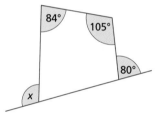

d)

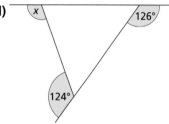

4 Find the size of the angles marked with letters. Give a geometric reason for each answer.

a)

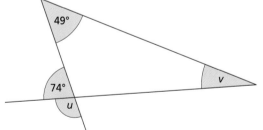

b)

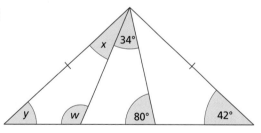

5 The diagram shows the kite *BCDE*.
ABC is a straight line and *AB = BE = BC*.

Find the sizes of angles *y* and *z*. Give geometric
reasons for your answers.

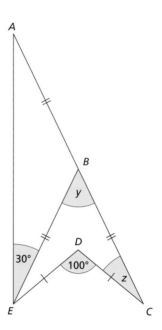

6 Vocabulary question Copy and complete the sentences using words from the box.

exterior angle	angles in a triangle	angles in a quadrilateral
interior angle	alternate	opposite
equal	corresponding angles	vertically opposite

An angle inside a shape is called an _____ . The angle formed outside a shape
between one side and the extension of the next side is called an _____ .

The exterior angle of a triangle is _____ to the sum of the _____ interior angles.

We can use angle rules and shape properties to find the value of a missing angle.
For example, _____ angles and _____ angles may be found on parallel lines
crossed by a transversal and are equal.

When any two lines intersect, two pairs of _____ equal angles are formed.

Angles on a straight line and _____ sum to 180°.

Angles around a point and _____ sum to 360°.

Thinking and working mathematically activity

In the triangle *ABC*, the angles *a*, *b* and *c* are all
interior angles.

What can you find out about the sum of angles *d*, *e* and *f*?

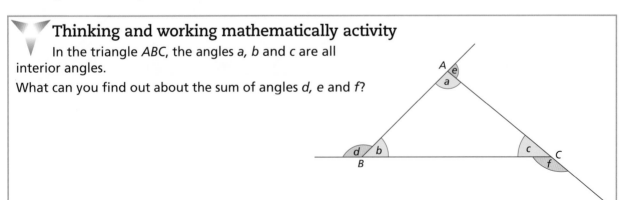

Key terms

A **bearing** is an angle measured clockwise from the north direction. It is always written with three digits, so a bearing of 5° from north would be written as 005° and a bearing of 50° from north would be written as 050°.

In the diagram, the clockwise direction (the way the hands of an analogue clock or watch go around) is shown by the arrow.

The bearing of the point P from the point O is written as 060° as it is 60° from north, measured clockwise. If you stood at O looking north, then you would have to turn through 60° clockwise before P was directly in sight.

The bearing Q from O is written as 240° as it is 240° from north measured clockwise: 180° from north to south and another 60° from south to Q.

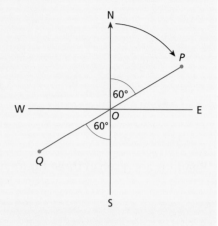

Worked example 4

a) What is the direction southeast written as a bearing?

The diagram shows two ships P and Q in the sea and a cliff.

b) What is the bearing of P from Q?

c) A lighthouse L, is on the edge of the cliff at a bearing of 320° from P. Mark the position of L onto the diagram.

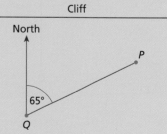

a) The angle of southeast from north is 90° + 45° = 135° The bearing of southeast is 135°	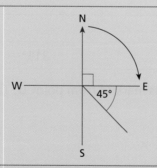

b) The bearing of P from Q is 065°

Draw the north line at the point where you are measuring the bearing from.

This is easiest to do on square paper but otherwise a set square can help you draw a north line parallel to the edge of the page.

Join P and Q and then measure the angle marked using a protractor.

It is 65° so as a bearing you need to add a leading zero to give the three digits needed for a bearing.

c)

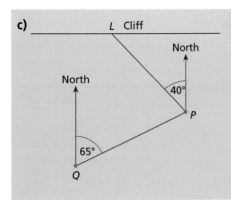

Start by drawing the north line at *P* as you are measuring the bearing of *L* from *P*.

320° from north is 40° less than 360° so measure 40° back from north to give the direction of *L* from *P*.

Draw in the line showing this direction. *L* is the point where this line crosses the edge of the cliff.

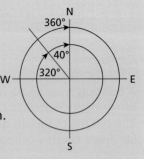

Did you know?

Bearings are given in three figure notation so that a value like 13° does not get confused with 130°.

Exercise 3

1 Copy and complete the table, using the compass to help you.

	Direction	Angle from north measured clockwise	Three figure bearing
a)	South		
b)			270°
c)		90°	
d)	Northeast		
e)		225°	
f)			315°

compass points

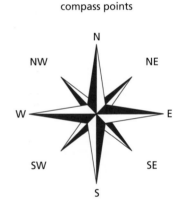

2 Measure the bearings of the points marked from point *O*. The north line at *O* is drawn for you.

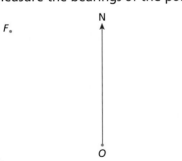

3 Draw a point O near the centre of a clean page. Draw lines 5 cm long from the point O to show the direction of the following bearings (you will need to draw the north line at O and label the lines with the letter for the part of the question):

a) 060°

b) 090°

c) 125°

d) 200°

e) 300°

4 A boat sails from the end of the pier P to a buoy Q and to a second buoy R and then back to P. A scale diagram of the journey is shown. Using your protractor measure:

a) the bearing of Q from P

b) the bearing of R from P

c) the bearing of R from Q

d) the bearing of Q from R.

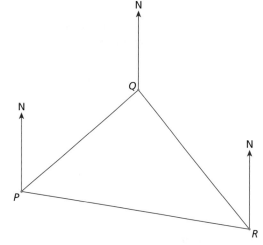

5 Jana and Alice look at the diagram in question 4. Jana says that the bearing of P from R looks to be about 280° as it is a bit more than three right angles. Alice says it looks more like 080° as it is almost 090°. Explain which of them is more likely to be correct and explain the mistake that the other one has made.

6 The bearing of P from Q is 065°. What is the bearing of Q from P?

 Thinking and working mathematically activity

• If you know the bearing from A to B, the bearing from B back to A is known as a back bearing.

• There is a rule you can always use to find the back bearing. Investigate this to find the rule.

Consolidation exercise

1 Write down whether the marked angles are vertically opposite angles, alternate angles or corresponding angles.

a)

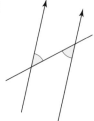

b)

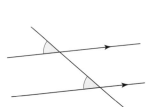

c)

2 Find the size of each angle marked with a letter.

a)

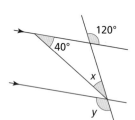

b)

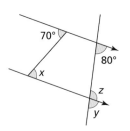

c)

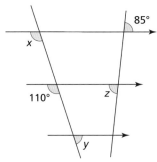

3 Find the size of the angles marked with letters in each diagram, stating the geometric reason for each step of your findings.

a)

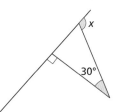

b)

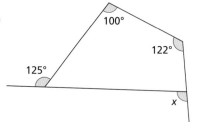

c)

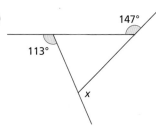

4 Ben said the lines *AB* and *BC* are parallel.
Is he correct?
Give reasons to support your answer.

5 Draw a point *O* near the centre of a clean page. Draw lines approximately 4 cm long from the point *O* to show the direction of the following bearings.
Show the North line from *O* and label your lines clearly.

a) 070°   **b)** 145°   **c)** 210°   **d)** 270°

6 A plane flies from airport *A* to base *B*.
   Then it flies from base *B* to camp *C*.
   Using your protractor, measure:

   a) the bearing of *B* from *A*

   b) the bearing of *C* from *B*

   c) the bearing of *A* from *C*

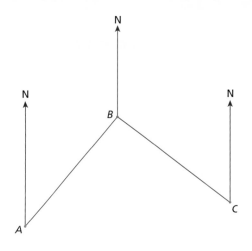

End of chapter reflection

You should know that...	You should be able to...	Such as...
When two lines intersect, the vertically opposite angles formed are equal. When a transversal crosses a pair of parallel lines, corresponding angles and alternative angles are equal.	Recognise vertically opposite angles. Spot corresponding and alternative angles.	Which angle (*a*, *b*, *c* or *d*) is the alternate angle to the angle marked yellow?
An exterior angle of a triangle is equal to the sum of the two opposite interior angles.	Understand a proof that shows an exterior angle of a triangle is equal to the sum of the two interior opposite angles. Understand a proof that shows the angle sum of a triangle is 180° Calculate angles in triangles and quadrilaterals, giving geometric reasons for your answers.	Complete the reasons for each step of the proof that an exterior angle of a triangle is equal to the sum of the two interior opposite sides. **1** $c = 180 - a - b$ (because ....) **2** $d = 180 - c$　(because .....) **3** So $d = 180 - (180 - a - b)$ 　　　(substituting for ....) **4** So $d = a + b$ (because ....)
Bearings are measured clockwise from north and are written in 3-digit notation.	Measure the bearing of one point from another point and write it in correct notation.	*P* is northeast of *Q*. What is the bearing of *P* from *Q*?

7 Place value, rounding and decimals

You will learn how to:

- Use knowledge of place value to multiply and divide integers and decimals by 0.1 and 0.01
- Round numbers to a given number of significant figures.
- Estimate and multiply decimals by integers and decimals.
- Estimate and divide decimals by numbers with one decimal place.

Starting point

Do you remember…

- how to multiply and divide integers and decimals by positive powers of 10?

  For example, write the values of $27 \div 1000$ and 0.32×10^2

- how to round numbers to a given number of decimal places?

  For example, round 6.16052 to 3 d.p.

- how to multiply positive and negative integers?

  For example, find $12 \times (-9)$ and $-65 \div 5$

- how to multiply and divide decimals by whole numbers, and how to estimate the answers?

  For example, estimate and then find 5.23×106 and $62.85 \div 5$

This will also be helpful when…

- you learn about negative indices, and learn to multiply and divide integers and decimals by 10 to the power of any positive or negative number
- you learn that when a number is rounded there are upper and lower limits for the number.
- you learn to divide by decimals that have more than one decimal place.

7.0 Getting started

Shino says, 'To multiply a whole number by 10, I add a zero to the number.
To multiply a decimal by 10, I move the decimal point one place to the right.'

Explain why Shino's rules work. You can use diagrams if that helps.

Write similar rules for dividing by 10. Explain how they work.

7.1 Multiplying and dividing by 0.1 and 0.01

Worked example 1

a) Multiply 15 by 0.1

b) Multiply 0.15 by 0.01

c) Divide 15 by 0.1

d) Divide 0.15 by 0.01

a) $15 \times 0.1 =$ $15 \div 10 = 1.5$	Multiplying by 0.1 is the same as dividing by 10	<table><tr><td>100</td><td>10</td><td>1</td><td>•</td><td>0.1</td><td>0.01</td><td>0.001</td></tr><tr><td></td><td></td><td>1</td><td>5</td><td>•</td><td></td><td></td></tr><tr><td></td><td></td><td></td><td>1</td><td>•</td><td>5</td><td></td></tr></table>
b) $0.15 \times 0.01 =$ $0.15 \div 100 = 0.0015$	Multiplying by 0.01 is the same as dividing by 100	<table><tr><td>100</td><td>10</td><td>1</td><td>•</td><td>0.1</td><td>0.01</td><td>0.001</td><td>0.0001</td></tr><tr><td></td><td></td><td>0</td><td>•</td><td>1</td><td>5</td><td></td><td></td></tr><tr><td></td><td></td><td>0</td><td>•</td><td>0</td><td>0</td><td>1</td><td>5</td></tr></table>
c) $15 \div 0.1 =$ $15 \times 10 = 150$	Dividing by 0.1 is the same as multiplying by 10	<table><tr><td>100</td><td>10</td><td>1</td></tr><tr><td></td><td>1</td><td>5</td></tr><tr><td>1</td><td>5</td><td>0</td></tr></table>
d) $0.15 \div 0.01 =$ $0.15 \times 100 = 15$	Dividing by 0.01 is the same as multiplying by 100	<table><tr><td>100</td><td>10</td><td>1</td><td>•</td><td>0.1</td><td>0.01</td></tr><tr><td></td><td></td><td>0</td><td>•</td><td>1</td><td>5</td></tr><tr><td></td><td>1</td><td>5</td><td>•</td><td></td><td></td></tr></table>

Exercise 1   1, 3–6

> **Think about**
>
> When does multiplying make a number smaller? When does dividing make a number larger?

1 Find:

     a) 3×0.1 **b)** $3 \div 0.1$ **c)** 3×0.01 **d)** $3 \div 0.01$

     e) $1.52 \div 0.1$ **f)** 15.9×0.01 **g)** 0.2×0.1 **h)** $0.025 \div 0.1$

     i) $1600 \div 0.1$ **j)** $0.281 \div 0.01$ **k)** 450×0.1 **l)** $0.0065 \div 0.01$

2 Use a calculator to find the value of:

     a) 1.4×0.1 **b)** 2.56×0.01 **c)** 0.128×0.1 **d)** 322×0.01

     e) $1.4 \div 0.01$ **f)** $2.56 \div 0.1$ **g)** $0.128 \div 0.01$ **h)** $322 \div 0.1$

3 Copy and complete these sentences.

    a) Multiplying by 0.1 is the same as dividing by ....

    b) Dividing by 0.1 is the same as multiplying by ....

    c) Multiplying by 0.01 is the same as dividing by ....

    d) Dividing by 0.01 is the same as multiplying by ....

4 Harpal has done two calculations. His working is below.

    In each calculation, what has Harpal done wrong? Write the correct answer.

a)

$$0.8 \times 0.1 = 8 \times 1$$
$$= 8$$

b)

$$0.5 \div 0.01 = 0.5 \div \frac{1}{100}$$
$$= \frac{1}{0.5} \times 100$$
$$= 2 \times 100$$
$$= 200$$

5 Copy and complete the following:

    a) $4 \times$ .... $= 0.4$ **b)** $1.2 \times$ .... $= 0.12$ **c)** $7.8 \times$ .... $= 0.078$

    d) $9 \div$ .... $= 90$ **e)** $6.2 \div$ .... $= 620$ **f)** .... $\div 0.01 = 4.85$

6 Copy and complete the following:

    a) $451 \times 0.1 =$ .... $\times 0.01$ **b)** $560 \div 0.1 =$ .... $\times 100$

    c) $38 \times 0.01 =$ .... $\times 10$ **d)** $6.2 \div 0.1 =$ .... $\div 0.01$

    e) $0.057 \div 0.01 =$ .... $\div 100$ **f)** $0.3 \times 0.1 =$ .... $\div 0.1$

> **Think about**
>
> If you divide a number by 0.01, is the result larger or smaller than when you divide it by 0.1?

▼ **Thinking and working mathematically activity**

    Copy and complete the table showing a pattern of multiplications of 75.

Under each arrow, write the operation that changes one number into the next number.

75×0.01	$75 \times$	75×1	75×10	$75 \times$ ....
	7.5	75		

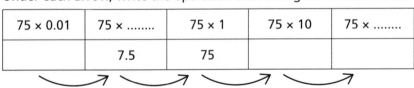

    $\times 10$

Copy and complete the table showing a pattern of divisions of 75.

Under each arrow, write the operation that changes one number into the next number.

$75 \div 0.01$	$75 \div$	$75 \div$	$75 \div$	$75 \div 100$
7500			7.5	

    $\div$ ....

Rewrite the ten calculations in three groups, with answers less than 75, equal to 75, and greater than 75.

Find any calculations that give the same answer. Try to explain this.

7.2 Rounding to significant figures

Key terms

The first **significant figure** (s.f.), or **significant digit**, of a number is the first non-zero digit.
For example, the first significant figure in 403 is 4, which has a value of 4 hundreds.
The first significant figure in 0.005208 is 5, which has a value of 5 thousandths.

Every digit after the first (or most) significant figure is also a significant figure. For example, in 403 the second significant figure is 0 and the third (or least) significant figure is 3. In 0.05208 there are four significant figures: 5, 2, 8, and the 0 between 2 and 8.

Rounding to a number of significant figures means rounding so that there are exactly that number of significant figures in the answer. For example, 0.05208 rounded to one significant figure is 0.05, and rounded to three significant figures is 0.0521

Did you know?

Zeroes at the end of a whole number are *sometimes* significant. It depends whether the number has been rounded. For example, 800 has three significant figures (3 s.f.) if it is exact, 2 s.f. if it has been rounded to the nearest 10, and only 1 s.f. if it has been rounded to the nearest 100. (You will not be expected to know about this.)

Worked example 2

a) Round 32 401 to 2 significant figures.

b) Round 0.03046 to 3 significant figures.

c) Round 0.0602 to 2 significant figures.

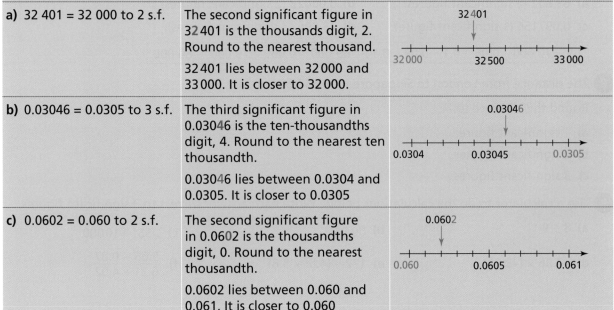

a) 32 401 = 32 000 to 2 s.f.	The second significant figure in 32 401 is the thousands digit, 2. Round to the nearest thousand. 32 401 lies between 32 000 and 33 000. It is closer to 32 000.	
b) 0.03046 = 0.0305 to 3 s.f.	The third significant figure in 0.03046 is the ten-thousandths digit, 4. Round to the nearest ten thousandth. 0.03046 lies between 0.0304 and 0.0305. It is closer to 0.0305	
c) 0.0602 = 0.060 to 2 s.f.	The second significant figure in 0.0602 is the thousandths digit, 0. Round to the nearest thousandth. 0.0602 lies between 0.060 and 0.061. It is closer to 0.060	

1 Write down the first significant figure in each number, and state what it represents.

(For example, in the number 568 the first significant figure is 5. It represents 5 hundreds.)

a) 76　　　　　b) 0.1058　　　　c) 0.00577　　　d) 81963

2 How many significant figures does each number have?

a) 3068　　　　b) 67067　　　　c) 127.93005　　d) 0.000244

3 Round 3.518 to:

a) 1 decimal place　b) 1 significant figure　c) 2 decimal places　d) 2 significant figures

4 Round each number to 1 significant figure.

a) 756　　　　　b) 742　　　　　c) 88111　　　　d) 94444

e) 11　　　　　f) 7.128　　　　g) 12.86　　　　h) 0.005023

5 Round each number to 2 significant figures.

a) 0.125　　　　b) 76008　　　　c) 426　　　　　d) 0.06789

e) 809　　　　　f) 803　　　　　g) 50001　　　　h) 0.002035

6 Round each number to 3 significant figures.

a) 5.2684　　　b) 1234　　　　　c) 0.06251　　　d) 32.353

e) 27350　　　　f) 0.004016　　　g) 0.4103　　　h) 3200

7 Round the numbers to the accuracy given. Choose your answers from the box below.

a) 69 802 (2 significant figures)　　　b) 0.06023 (2 significant figures)

c) 0.097156 (1 significant figure)　　　d) 78 996 (3 significant figures)

69　70 000　0.06　0.1　0.09　0.10　69 000　0.060　79 000　80 000

8 The distance from London to Singapore is 10 848.68 km.

Round this distance to:

a) 1 significant figure

b) 2 significant figures

c) 3 significant figures

9 Use a calculator to do the calculations below. Write each answer correct to 3 significant figures.

a) $8 \div 9$　　　　　　　　b) $\sqrt[3]{9.86}$　　　　　　　c) $\sqrt{55^3 + 10000}$

d) $-4.86 \times (-3.92)$　　　　e) $172 - (4.6^2 \times 5.8)$　　　f) $\dfrac{5.65 - 0.87}{6.31 + 4.02}$

10 Suman says that 3489 rounded to 1 significant figure is 4000. His working is below.

> 3489 to 3 significant figures is 3490
>
> 3490 to 2 significant figures is 3500
>
> 3500 to 1 significant figure is 4000

Explain why Suman is incorrect.

Write the correct answer.

11 Goran says that 999 rounded to 2 significant figures is 990. Mel says it is 1000.

Which answer is correct? Explain your answer.

12 Write a number that is the same whether it is:

a) rounded to 1 decimal place or 1 significant figure

b) rounded to 2 decimal places or 1 significant figure

c) rounded to 1 decimal place or 2 significant figures.

> **Discuss**
>
> When is the digit 0 a significant figure, and when is it not a significant figure?

Thinking and working mathematically activity

Technology question Use a spreadsheet for this investigation and present your calculations in an organised way.

A whole number is divided by 9. The answer, rounded to one significant figure, is 8.
Find the largest and smallest possible values for the original number.

A whole number is divided by 9. The answer, rounded to one significant figure, is 7.
Find the largest and smallest possible values for the original number.

Repeat for division by 9 with answers 6, 5 and 4 (to 1 s.f.)

Describe any pattern you can see. Try to explain it.

In a similar way, investigate division by 8.

7.3 Multiplying and dividing with integers and decimals

Worked example 3

Estimate and then calculate:

a) 0.67 × (−9)

b) 0.34 × 0.84

a) 0.67 × (−9) ≈ 0.7 × (−9)	To estimate, write each number correct to 1 significant figure.
7 × (−9) = −63 ÷ 10 ⟲ ⟳ ÷ 10 0.7 × (−9) = −6.3	Ignore the decimal point and find 7 × (−9). 7 is 10 times bigger than 0.7, so 63 is 10 times too big. The estimated answer is −6.3
$\begin{array}{r} 6\,7 \\ \times\ \ 9 \\ \hline 60^{6}3 \end{array}$	To find the exact answer, first ignore the decimal point and find 67 × (−9) 67 × 9 = 603, so 67 × (−9) = −603

> **Did you know?**
>
> To estimate the answer to a calculation, you can round each number to 1 significant figure and then find the result.

$0.67 \times (-9) = -6.03$	Use the estimated answer, -6.3, to see where to place the decimal point.
b) $0.34 \times 0.84 \approx 0.3 \times 0.8$ $3 \times 8 = 24$ $\div 100$ ⟳ $\div 100$ $0.3 \times 0.8 = 0.24$ $\begin{array}{r} 34 \\ \times\ 84 \\ \hline 13\overset{1}{6} \\ 27\overset{3}{2}0 \\ \hline 2856 \end{array}$	To estimate, write each number correct to 1 significant figure. Ignore the decimal points and find 3×8 3 is 10 times bigger than 0.3, and 8 is 10 times bigger than 0.8 So, 24 is $10 \times 10 = 100$ times too big. The estimated answer is 0.24 To find the exact answer, first find 34×84
$0.34 \times 0.84 = 0.2856$	Use the estimated answer, 0.24, to see where to place the decimal point.

Worked example 4

a) Find: $-6 \div 0.4$

b) Estimate and then calculate $2.76 \div 0.8$

a) $-6 \div 0.4 = \dfrac{-6}{0.4}$	Write the division as a fraction.
$\dfrac{-6}{0.4} = \dfrac{-60}{4} = -60 \div 4$	Make the divisor a whole number without changing the value of the fraction: multiply the numerator and denominator by 10. This shows that $-6 \div 0.4 = -60 \div 4$
$\begin{array}{r} 15 \\ 4\overline{)6\overset{2}{0}} \end{array}$ $-6 \div 0.4 = -15$	$60 \div 4 = 15$, so $-60 \div 4 = -15$
b) $2.76 \div 0.8 \approx 3 \div 0.8$ $\qquad\qquad = 30 \div 8$ $\qquad\qquad \approx 3$	To estimate, write each number correct to 1 significant figure. You can ignore any remainder in the division.
$2.76 \div 0.8 = \dfrac{2.76}{0.8}$	Write the division as a fraction.
$\dfrac{2.76}{0.8} = \dfrac{27.6}{8} = 27.6 \div 8$	Make the denominator a whole number. You can do this by multiplying the numerator and denominator by 10. This shows that $2.76 \div 0.8 = 27.6 \div 8$
$\begin{array}{r} 3.45 \\ 8\overline{)27.\overset{3}{6}\overset{4}{0}} \end{array}$ $2.76 \div 0.8 = 3.45$	Find $27.6 \div 8$

1 a) Copy and complete:

0.2 × _____ = 0.8

_____ × 0.4 = 0.08

0.2 × 0.04 = ____

b) Find:

i) 2 × 0.4   **ii)** 0.2 × 4   **iii)** 0.02 × 0.4   **iv)** 0.02 × 0.04

2 Find:

a) −4 × 0.5   **b)** −7 × 0.6   **c)** 1.4 × 3   **d)** 0.25 × (−6)

3 Below is a list of numbers.

1.302   2.752   0.2752   4.29   5.772   9.46

Each number is the answer to one of the calculations below.

Use estimation to choose the correct answer to each calculation.

a) 0.32 × 0.86   **b)** 1.4 × 0.93   **c)** 11.1 × 0.52

d) 0.8256 ÷ 0.3   **e)** 5.676 ÷ 0.6   **f)** 3.003 ÷ 0.7

4 Estimate and then calculate:

a) 3.1 × 1.1   **b)** 5.4 × 0.32   **c)** 0.45 × 0.39   **d)** 1.8 × 0.07

e) 0.28 × 6.3   **f)** 0.08 × 0.44   **g)** 0.13 × 0.04   **h)** 0.73 × 0.96

5 426 × (−5) = −2130. Use this to write the value of:

a) 4.26 × (−5)   **b)** −4260 × 0.5   **c)** 42.6 × 0.05   **d)** 0.426 × 0.005

6 Jackie sells lemonade for $0.60 per cup. She sells all of her lemonade for $22.80.

Estimate and then calculate how many cups of lemonade she sells.

7 Find:

a) 6 ÷ 0.5   **b)** −12 ÷ 0.6   **c)** 8 ÷ 0.2   **d)** −20 ÷ 0.4

8 Estimate and then calculate:

a) 52 ÷ 0.4   **b)** 22 ÷ 0.8   **c)** −180 ÷ 0.8   **d)** −44 ÷ 0.2

e) 4.3 ÷ 0.2   **f)** 5.7 ÷ 0.5   **g)** 16.2 ÷ 0.3   **h)** 1.32 ÷ 0.6

9 Estimate and then calculate each answer.

Round your calculated answers to 3 significant figures.

a) −46 ÷ 0.3   **b)** 88.3 ÷ 0.5   **c)** 12.32 ÷ 0.9   **d)** −77 ÷ 0.8

10 Explain the mistakes in each calculation and then correct them.

a) $6.9 \times 0.7 = 48.3$

$$\begin{array}{r} 6.9 \\ \times 0.7 \\ \hline 6\ 3 \\ 420 \\ \hline 48.3 \end{array}$$

b) $371 \div 0.7 = 53$

$$\begin{array}{r} 53 \\ 7\overline{)37^21} \end{array}$$

Thinking and working mathematically activity

Find the missing digits in each calculation. Afterwards, discuss your methods with other students.

$0.8\,\square \div 0.3 = \square.7$

$6.\square 4 \div 0.3 = 21.\square$

$0.9\,\square \div 0.4 = \square.3$

$7.\square 3 \div 0.9 = \square.7$

$\square.8 \times 0.9 = 5.2\,\square$

$\square.6 \times 1.2 = \square.32$

How many possible solutions are there to each problem below? How do you know if you have found them all?

$1\,\square.28 \div 0.4 = \square 5.7$

$5.\square \div 0.2 = 2\,\square$

$0.7\,\square \times 0.5 = 0.3\,\square 5$

Consolidation exercise

1 Arrange the cards to make two correct number statements.

Use each card once only.

×	÷	=	=	0.23	0.23	2.3	23	0.1	0.01

2 m is a number greater than zero. Place the calculations in order of the size of their answer, smallest first.

$m \div 0.1$ $m \times 0.1$ $m \times 0.01$ $m \div 0.01$

3 Phil says, 'I round a number to 2 significant figures, and the answer is 100.'

Enrique says, 'That is not possible, because 100 has only 1 significant figure.'

Is Enrique correct? Explain your answer.

4 Which of these numbers have exactly 3 significant figures?

0.089   12.6   0.40021   0.508   1.508

5 Write down a number with 3 decimal places that would be:

a) 4, when rounded to 1 significant figure; and 4.2, when rounded to 2 significant figures.

b) 4.2, when rounded to 2 significant figures; and 4.25, when rounded to 3 significant figures.

6 Strawberries cost $5.20 per kg.

For each amount below, estimate and then calculate the cost.

a) 0.2 kg of strawberries   b) 0.05 kg of strawberries   c) 1.8 kg of strawberries

7 A painter needs to paint a white line along a road 16.8 km long. With one tin of paint she can paint a line 0.6 km long.

Estimate and then calculate how many tins of paint she needs.

8 Laila puts 9.6 kg of potatoes into bags. Each bag contains 0.8 kg of potatoes.

Estimate and then calculate how many bags she makes.

9 In each pair, which is bigger? How many times bigger?

a) $19 \div 0.01$ and 19×0.01

b) $5.6 \div 0.7$ and $56 \div 700$

10 Vocabulary question Explain how to use the calculation $58 \times 0.24 = 13.92$ to find the answer to 5.8×240

You must use the words: multiply (or multiplied), divide (or divided), ten, one hundred, one thousand, greater than.

End of chapter reflection

You should know that...	You should be able to...	Such as...
Multiplying by 0.1 is equivalent to dividing by 10 Multiplying by 0.01 is equivalent to dividing by 100	Multiply integers and decimals by 0.1 or 0.01	Find: a) 17×0.1 b) 3.7×0.01
Dividing by 0.1 is equivalent to multiplying by 10 Dividing by 0.01 is equivalent to multiplying by 100	Divide integers and decimals by 0.1 or 0.01	Find: a) $0.432 \div 0.1$ b) $8 \div 0.01$
The first significant digit in a number is the first non-zero digit from the left. Every digit after this is significant.	Round integers and decimals to a given number of significant figures.	Round: a) 56 223 to 1 significant figure b) 0.0503 to 2 significant figures

You can estimate the answer to a calculation by rounding each number to 1 significant figure.	Estimate the answer to a multiplication or division by rounding each number to 1 significant figure.	Estimate: **a)** $7.8 \times (-45)$ **b)** 0.12×0.9 **c)** $5.64 \div 0.3$
To multiply or divide a decimal by a negative integer, you can use the rules about multiplying positive and negative numbers.	Multiply or divide a decimal by a negative integer.	Calculate: **a)** -9×0.86 **b)** $8.75 \div (-7)$
To multiply two decimals, you can: • ignore the decimal points and multiply as integers • then write the decimal point at the correct place in the answer.	Multiply two decimals.	Calculate 1.7×0.023
To divide by a decimal, you can: • multiply the dividend and divisor by the same number, to make the divisor an integer • then use a written method of division.	Divide an integer or a decimal by a decimal.	Calculate $2.71 \div 0.8$

8

Presenting and interpreting data 1

You will learn how to:

- Record, organise and represent categorical, discrete and continuous data. Choose and explain which representation to use in a given situation:
  - ○ tally charts, frequency tables and two-way tables
  - ○ dual and compound bar charts
  - ○ frequency diagrams for continuous data
  - ○ stem-and-leaf diagrams
- Interpret data, identifying patterns, trends and relationships, within and between data sets, to answer statistical questions. Discuss conclusions, considering the sources of variation, including sampling, and check predictions.

Starting point

Do you remember...

- what the symbols < and ≤ mean?

  For example, are these statements true or false? **a)** $3.403 < 3.4$ **b)** $\frac{24}{5} \leq 4.8$

- how to draw and interpret frequency diagrams?

  For example, in the diagram below, how many people were at least 1.6 metres in height?

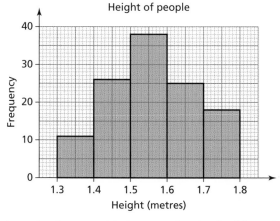

- how to draw and interpret simple dual bar charts and compound bar charts?

  For example, in the bar chart below, how many students aged 11 years chose Chess club?

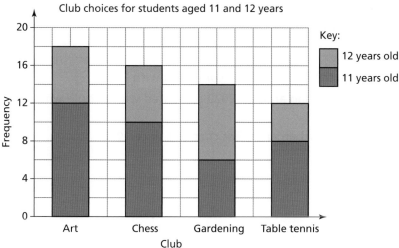

- how to find the mean, median, mode and range?

  For example, find the mean, median, mode and range for these data:

  | 45 | 52 | 78 | 65 | 39 | 72 | 55 | 64 |
  | 46 | 62 | 59 | 52 | 47 | 33 | 61 | 47 |

This will also be helpful when…

- you find the mean from grouped frequency tables
- you draw and interpret back-to-back stem-and-leaf diagrams.

8.0 Getting started

Real data question The graph shows changes in the world's population between 1900 and 2020, as well as the number of people living in urban and rural areas.

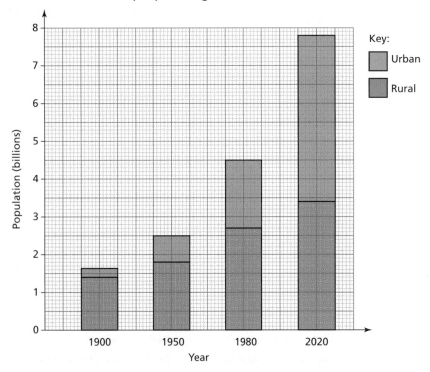

Source: Hannah Ritchie and Max Roser (2018) - "Urbanization". Published online at OurWorldInData. org. Retrieved from: 'https://ourworldindata.org/urbanization' [Online Resource].

Use the graph to find:

- how many times more people live in the world in 2020 compared with 1900.
- how many times more people lived in urban areas in 2020 compared with people who lived in urban areas in 1900.
- how the percentage of people living in urban areas has changed since 1900.

Make some conclusions from the graph.

8.1 Frequency tables and diagrams

Key terms

Class intervals are used to group continuous data in a frequency table. You can use class intervals with equal widths or class intervals with unequal widths. The class interval $10 \leq x < 20$ represents values of x from 10 up to (but not including) 20.

The **modal class** is the class interval that is most frequent.

Worked example 1

Here are the times (in minutes) that 24 tennis games lasted.

| 102 | 126 | 216 | 104 | 66 | 93 | 129 | 186 | 54 | 73 | 194 | 138 |
| 98 | 77 | 145 | 90 | 238 | 55 | 87 | 165 | 181 | 94 | 110 | 176 |

a) Design a frequency table suitable for recording the data.

b) Complete the frequency table.

c) Write down the modal class.

d) Draw a frequency diagram to show the information.

a)

Time, t (min)	Tally	Frequency
$40 \leq t < 80$		
$80 \leq t < 120$		
$120 \leq t < 160$		
$160 \leq t < 200$		
$200 \leq t < 240$		

Identify the smallest and largest values:

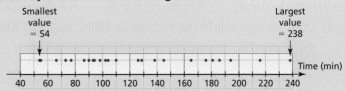

Choose suitable class intervals that cover the data – you should aim for between four and eight classes. The classes should be convenient to work with and are usually chosen with equal widths.

b)

Time, t (min)	Tally	Frequency
$40 \leq t < 80$	卌	5
$80 \leq t < 120$	卌 |||	8
$120 \leq t < 160$	||||	4
$160 \leq t < 200$	卌	5
$200 \leq t < 240$	||	2

Take each data value in turn and place a tally mark in the corresponding row.

Remember that the fifth tally mark in any class is drawn through the previous four so that the tally marks are grouped in sets of five.

Count up the tallies in each class interval and complete the frequency column.

c) The modal class is $80 \leq t < 120$

The modal class is the class interval with the highest frequency.

The highest frequency is 8. This corresponds to the class interval $80 \leq t < 120$.

d)

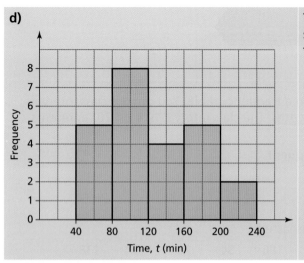

The data are continuous so the frequency diagram should have a continuous scale. The bars should be touching. Remember to label the axes.

Exercise 1

1 These are the heights of 12 trees in an orchard:

3.5 m	2.9 m	5.1m	7.9 m	6.0 m	9.1 m
5.5 m	1.2 m	4.5 m	8.3 m	7.7 m	6.3 m

a) Copy and complete the frequency table for these data.

Height, h (m)	Tally	Frequency
$0 < h \leq 2$		
$2 < h \leq 4$		
$4 < h \leq 6$		
$6 < h \leq 8$		
$8 < h \leq 10$		

b) How many trees are 4 m or shorter in height?

2 a) Write down the missing class interval in this frequency table.

b) Write down the missing value in the frequency column.

Mass, m (grams)	Frequency
$0 < m \leq 5$	4
$5 < m \leq 10$	7
	6
$15 < m \leq 20$	8
$20 < m \leq 25$	
$25 < m \leq 30$	11
Total	**41**

3 The diameters of 14 tomatoes are:

| 41 mm | 49 mm | 62 mm | 46 mm | 58 mm | 61 mm | 57 mm |
| 55 mm | 58 mm | 54 mm | 60 mm | 59 mm | 56 mm | 56 mm |

a) Copy and complete the frequency table for these data using equal class intervals.

Diameter, d (mm)	Tally	Frequency
$40 < d \leq 45$		
$45 < d \leq 50$		
$50 < d \leq 55$		
$55 < d \leq 60$		
$60 < d \leq 65$		

b) How many tomatoes had a diameter of more than 50 mm?

4 Design and fill in a frequency table for each of these sets of data.

a)
Mass (grams)				
28.4	27.5	29.1	26.3	27.8
28.6	27.2	27.5	28.3	25.7
29.3	26.2	27.3	26.9	28.5

b)
Number of people					
81	75	66	62	72	78
68	74	64	82	70	64
72	79	77	76	72	69

c)
Number of parcels					
381	291	652	335	376	618
407	525	493	380	671	428
576	493	465	266	526	398
673	552	518	470	601	374

d)
Temperature (°C)					
27.3	28.4	32.4	11.4	32.4	14.2	19.6
17.4	32.7	29.0	13.2	17.4	37.8	29.1
26.1	22.2	14.5	19.7	33.1	27.3	15.2
20.7	31.2	29.3	30.2	26.0	17.1	29.3

> **Tip**
>
> In each part of question 4, look to see whether the data are discrete or continuous – this will affect how you present the class intervals in your frequency table.

5 The frequency table below shows the speed, in km/h, of some cars as they passed a school between 2:55 p.m. and 3:00 p.m. The school is in a 40 km/h zone.

Speed, s (km/h)	Frequency
$0 < s \leq 10$	1
$10 < s \leq 20$	5
$20 < s \leq 30$	3
$30 < s \leq 40$	10
$40 < s \leq 50$	5
$50 < s \leq 60$	4

a) How many cars were recorded?

b) How many cars were travelling 30 km/h or slower?

c) How many cars drove above the speed limit?

d) Joe says that $\frac{1}{4}$ of the cars drove above the speed limit. Is Joe correct? Give a reason for your answer.

> **Think about**
>
> If $5 < x \leq 10$,
>
> what is the largest value x could be?
>
> What is the smallest value x could be?

6 Draw a frequency diagram to show each set of data.

a)

Length of lorry, x (metres)	Number of lorries
$8 \leq x < 10$	9
$10 \leq x < 12$	16
$12 \leq x < 14$	8
$14 \leq x < 16$	7

> **Tip**
>
> Think carefully about whether each table shows discrete or continuous data.

b)

Number of insects	Frequency
0 – 4	6
5 – 9	11
10 – 14	14
15 – 19	9

c)

Temperature, t (°C)	Number of days
$20 \leq t < 21$	11
$21 \leq t < 22$	4
$22 \leq t < 23$	9
$23 \leq t < 24$	6
$24 \leq t < 25$	1

7 Donna records the heights of 50 trees.

a) Draw a frequency diagram to show the information.

b) Find how many of the trees have a height of at least 14 metres.

c) Write down the modal class.

d) Calculate the percentage of the trees that are under 10 metres tall.

Height, h (m)	Number of trees
$2 \leq h < 6$	6
$6 \leq h < 10$	10
$10 \leq h < 14$	15
$14 \leq h < 18$	12
$18 \leq h < 22$	7

8 The frequency diagram shows the times taken by a group of students to run a race.

Decide if each of the following statements is true or false. Give a reason for each answer.

A 27 students took less than 50 seconds to run the race.

B The modal class is 60 – 65 seconds.

C 60 students took part in the race.

D 25% of students took more than one minute to run the race.

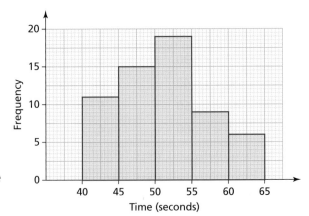

9 The frequency diagram shows the lengths for a sample of jellyfish.

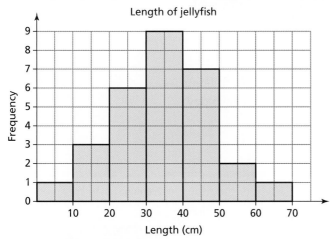

Length of jellyfish

a) Copy and complete this frequency table to show the information in the diagram. The first row has been done for you.

Length, L (cm)	Frequency
$0 \leq L < 10$	1
...	...

b) Write down the modal class.

c) Calculate the percentage of jellyfish in the sample that are longer than 50 cm. Give your answer to 1 decimal place.

Thinking and working mathematically activity

Real data question The diagrams show the percentage of the world's population in different age groups in 1960 and in 2020.

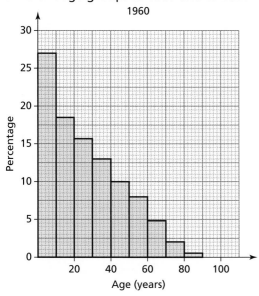

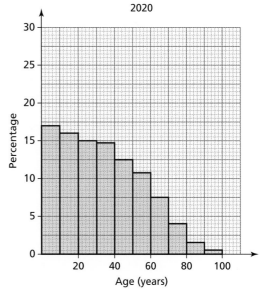

Source: © December 2019 by PopulationPyramid.net

Make some conclusions from the diagrams. Suggest possible reasons for the differences.

Draw a possible diagram to predict what the age distribution for the world might be like 60 years from now. Give reasons for your frequency diagram.

Worked example 2

A museum exhibition took place on four days, Monday to Thursday.

The compound bar chart shows the number of visitors on each day.

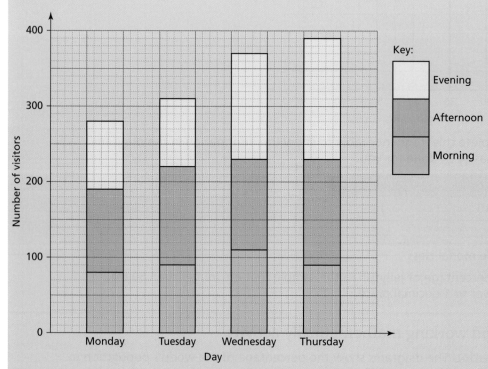

a) Comment on how the number of visitors to the exhibition changed over the four days.

b) Find the total number of visitors attending over the four days.

c) Find how many visitors went to the exhibition on Tuesday afternoon.

a) The number of visitors has increased each day.	The total height of the bars represents the total number of visitors each day. These heights are increasing from one day to the next.	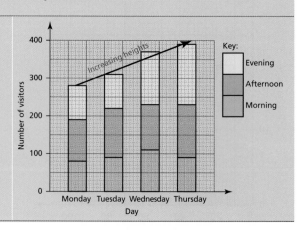

b)

Monday: 280 visitors

Tuesday: 310 visitors

Wednesday: 370 visitors

Thursday: 390 visitors

Total: 1350 visitors

Find the number of visitors who attended on each day.

Read off the figures from the top of the scale. Notice that each small square represents 10 visitors.

Add the figures for each day together.

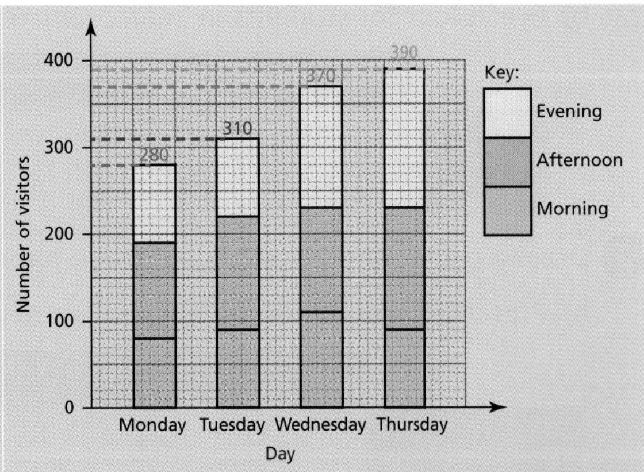

c) Number of visitors on Tuesday afternoon is

$220 - 90 = 130$

The number of visitors for Tuesday afternoon is represented by the height of the pink bar.

The vertical axis reading at the bottom of the pink bar is 90.

The vertical axis reading at the top of the pink bar is 220.

Subtract these values to get the number of visitors for the afternoon.

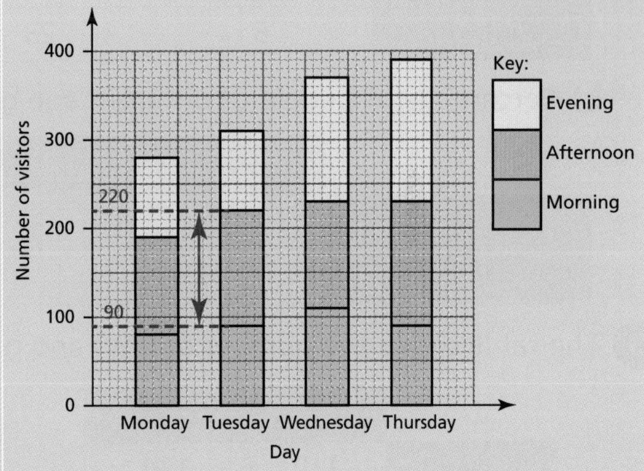

Did you know?

Compound bar charts are useful as they show lots of information in one graph. The compound bar chart in the example:

- shows information about the total number of people attending on each day
- allows you to compare the number of people attending at different times of the day.

Exercise 2

1 Draw a dual bar chart to show the following sets of data. Draw your diagrams on graph paper.

a) Vehicles passing a school in the morning and in the afternoon:

	Car	Motorbike	Vans/lorries
Morning	57	17	20
Afternoon	43	11	29

b) Eye colour for students in Year 7 and Year 8:

	Brown	Blue	Amber	Green
Year 7	86	24	4	0
Year 8	92	16	10	6

2 Draw a compound bar chart on graph paper to show the following sets of data.

a) Preferred activity for members of an activity club:

	Walking	Kayaking	Climbing	Biking
Men	17	8	32	16
Women	13	17	21	28
Children	5	23	17	15

b) Percentage of people of different age groups who support the building of a new skate park:

	Under 20	20–40	Over 40
Support	94%	85%	69%
Do not support	6%	15%	31%

3 The table shows the number of boys and girls present and absent one day from school.

	Boys	Girls
Present	110	144
Absent	35	16

a) Draw a compound bar chart to show the information, with one bar for boys and one for girls.

b) Make one comment about the numbers of children absent.

4 The dual bar chart shows the number of small and large sofas made by a factory between January and April.

a) Write down how many large sofas were made in February.

b) Calculate the total number of sofas made in January.

c) Calculate how many more small sofas than large sofas were made in March.

d) Find which month the factory made the most sofas.

e) Make some comments about what the chart shows about the sofas made.

f) Give a reason why a dual bar chart is an appropriate diagram to show the information.

5 A college offers online language courses in English, Mandarin and Arabic. Students are awarded either a Gold, a Silver or a Bronze certificate for the course at the end of the year.

The compound bar chart shows the results of students taking each course one year.

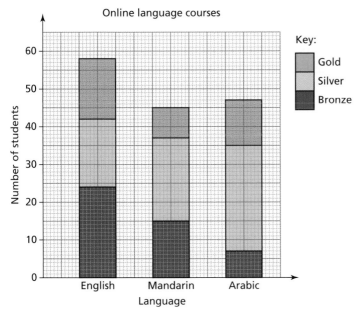

a) Write down the number of students who studied Arabic.

b) Calculate how many more students studied English than studied Mandarin.

c) Find the number of students who were awarded a Gold certificate in English.

d) Find the fraction of Mandarin students who were awarded a Bronze certificate.
   Give your answer in its simplest form.

6 **Real data question** The compound bar chart shows the percentage of people (who are available for work) who were employed and unemployed in four South American countries in 2019.

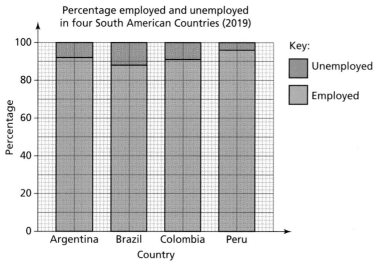

Percentage employed and unemployed in four South American Countries (2019)

Key:
Unemployed
Employed

Source: UNdata

a) Write down the percentage of people who were unemployed in Argentina in 2019.

b) Make some comparisons about the percentage of unemployed people in these countries.

7 Harry asks a random sample of students from his school what type of reading they prefer.

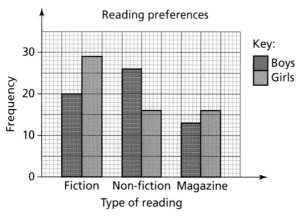

Reading preferences

Key:
Boys
Girls

a) Calculate the total number of students in Harry's sample.

b) How many more girls than boys preferred fiction books?

c) Copy and complete this table to show the information.

	Fiction	Non-fiction	Magazine
Boys			
Girls			

d) Harry makes the following conclusion:

Non-fiction is the preferred type of reading for about one third of students in the school.

Comment on Harry's conclusion.

Thinking and working mathematically activity

A club has junior and senior members.

The graph shows how the number of each type of member has changed between 2016 and 2019.

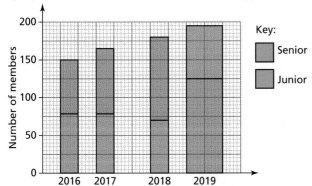

- Comment on the presentation of the graph. Draw the graph correctly.
- Draw a different type of graph to show the same information.
  Discuss which graph shows the information most clearly.

8.3 Stem-and-leaf diagrams

Key terms

Stem-and-leaf diagrams are a way of displaying data that retains all the actual data values.

A stem-and-leaf diagram to show the ages of 16 people

Stem	Leaf
1	2 5
2	0 4 7
3	1 1 4 6 9
4	3 8
5	0 2 5 5

Key: 5 | 2 = 52 years old

Note that:
- the leaf values are written in order of size.
- the leaf values are written in columns.
- a key is given to explain what each value represents.

The numbers in a stem-and-leaf diagram are in order, so the **median** value can easily be found.

Worked example 3

A shop recorded the number of jars of coffee it sold each day.

Here are the numbers for the last 20 days.

| 46 | 52 | 71 | 67 | 55 | 72 | 63 | 60 | 48 | 54 |
| 49 | 61 | 56 | 58 | 52 | 64 | 48 | 45 | 65 | 57 |

a) Draw a stem-and-leaf diagram to show these data.

b) Use your diagram to find the median number of jars of coffee the shop sold each day.

a)
Stem	Leaf
4	5 6 8 8 9
5	2 2 4 5 6 7 8
6	0 1 3 4 5 7
7	1 2

Key: 4 | 5 = 45 jars

Start by writing the data values in increasing order.
Split each number into two parts, the stem (here the tens digit) and the leaf (here the units digit).

Make sure that in the stem-and-leaf diagram:

- you write the leaf values in order

- you write the leaf values in columns

- you give a key to explain what each value represents.

Ordered list of data values:

45 46 48 48 49

52 52 54 55 56 57 58

60 61 63 64 65 67

71 72

b) The 10th value is 56.
The 11th value is 57.

The median is halfway between the 10th and 11th values.

So, the median is 56.5 jars of coffee.

The median is the value that is halfway through the data.

There are 20 data values, so the median will be halfway between the 10th and the 11th values.

Stem	Leaf
4	5 6 8 8 9
5	2 2 4 5 (6 7) 8
6	0 1 3 4 5 7
7	1 2

Key: 4 | 5 = 45 jars

Did you know?

Stem-and-leaf diagrams became a popular way to display data in the 1980s because they could easily be drawn on early computers.

Worked example 4

A football coach wants to investigate the fitness of his players.

He plans to make every player run a race.

The coach has this hypothesis:

Less than half of the players will be able to run the race in under 1 minute.

The times of each of the 32 players in the race are shown in the stem-and-leaf diagram.

Investigate the coach's hypothesis.

Stem	Leaf
4	5 6 7 8 9 9
5	0 1 2 2 4 5 6 7 8 9 9
6	1 1 2 3 3 3 4 5 5 6 7 8
7	0 2 5

Key: 4 | 5 = 45 seconds

17 players completed the race in less than 1 minute. Half of the 32 players is 16. So the coach's hypothesis is not true. More than half of the players were able to run the race in under 1 minute.	First find how many players completed the race in under 60 seconds. Then make a conclusion by referring to the coach's hypothesis.	Stem | Leaf 4 | 5 6 7 8 9 9 5 | 0 1 2 2 4 5 6 7 8 9 9 6 | 1 1 2 3 3 3 4 5 5 6 7 8 7 | 0 2 5 Key: 4 | 5 = 45 seconds The times in red are all less than 1 minute = 60 seconds.

Exercise 3

1 Draw a stem-and-leaf diagram to show each set of data. Remember to include a key.

a)

56	57	59	61	64	65	67	69
70	75	77	77	79	81	82	

b)

19	24	45	35	53	26	38
27	36	34	52	35	33	41

c)

13.1	12.5	14.7	12.8	13.6	13.4
15.2	12.5	13.4	14.3	14.8	13.9

Tip

In 1c, the whole number part of each value is the stem and the decimal part is the leaf.

Discuss

How could you find the median for each data set in question 1?

2 The stem-and-leaf diagram shows the masses of 25 apples.

Stem	Leaf
9	2 4 5 6
10	0 2 4 5 5 8 8
11	1 1 4 4 4 7
12	2 3 5 6 8
13	1 4 9

Key: 9 | 2 = 92 grams

a) How many apples have a mass less than 100 grams?

b) Find the fraction of the apples that have a mass between 120 grams and 130 grams.

c) Write down the mass of the heaviest apple.

d) Find the range of masses of the apples.

e) Write down the mass that is the mode.

f) Find the median mass of the apples.

3 Jacques has 15 apple trees. He records the number of apples he picked from each tree last summer.

14	9	21	16	28	33	5	41
19	34	25	30	29	17	21	

    a) Show this information in a stem-and-leaf diagram.

    b) Use your diagram to find the median number of apples he picked from his trees.

4 The stem-and-leaf diagram shows the lengths of 30 insects.

    a) How many insects were 4.5 cm long?

    b) Find the percentage of the insects that have a length greater than 3.8 cm.

    c) Find the range of the lengths of the insects.

    d) Find the median length of the insects.

Stem	Leaf
1	2 5 6 8 9
2	1 3 5 6 7 8
3	1 1 2 3 5 6 7 9
4	1 5 5 5 6 7
5	0 4 5 5 8

Key: 1 | 2 = 1.2 cm

5 Greg investigates the number of students that are learning a musical instrument in each school in his area. The stem-and-leaf diagram shows his results.

Make two conclusions about the number of students learning instruments in the 24 schools.

Stem	Leaf
0	0 7
1	2 3 5 5 9
2	0 1 2 4 5 6 7
3	1 2 6 7 8 9
4	1 3 5
5	2

Key: 1 | 2 = 12 students

6 Silva is investigating this hypothesis:

The median height of adult women is greater than 155 cm.

She records the height of a random sample of 35 women.

Make a conclusion about Silva's hypothesis. Use the data to explain your answer.

Stem	Leaf
13	6 9
14	3 4 6 6
15	2 2 3 4 6 7 8 9
16	0 1 1 2 4 5 5 6 7 8
17	1 3 5 6 6 8
18	2 3 4 5
19	1

Key: 13 | 6 = 136 cm

7 **Real data question** The times (in seconds) recorded in the Men's heats of the 100 metres in the Rio Paralympic Games for wheelchair athletes are:

14.6	14.8	14.8	15.4	16.8	18.8	18.8
14.1	14.3	14.4	14.6	15.1	15.6	19.3
14.0	14.4	14.7	15.3	15.3	15.4	17.5

Source: International Paralympic Committee, https://www.paralympic.org/

a) Draw a stem-and-leaf diagram to show the times.

b) Find the median of these times.

c) The World Record for this event is 13.8 seconds. Find the fraction of athletes in these heats that finished within 2 seconds of this world record time.

Thinking and working mathematically activity

Real data question The data show the mean daily temperatures (in °C) for two cities in Japan – Tokyo and Sapporo. Data are given for the month of April for years between 2005 and 2019.

Tokyo

12	4				
13	6	6	7		
14	5	5	5	7	7
15	0	1	2	4	7
16					
17	0				

Key 12 | 4 = 12.4°C

Sapporo

5	2	5		
6	2	3	3	9
7	0	3	7	7	8
8	0	2	7	
9	4			

Key 5 | 2 = 5.2°C

Source: Japan Meteorological Agency website (https://www.data.jma.go.jp/obd/stats/etrn/view/monthly_s3_en.php?block_no=47412&view=1).

Working with a partner, write down some statements (which could be true or could be false) about the diagrams.

Swap your statements with another pair. Sort the statements you are given according to whether they are true or false.

Think about

Compare the data for the Japanese cities with the temperatures in April where you live.

Consolidation exercise

1 Draw an appropriate diagram for the data shown in each table.

a)

Number of students	0–3	4–7	8–11	12–15	16–19	20–23
Number of days	66	70	55	22	12	10

b)

Height, h (m)	$1.3 < h \le 1.4$	$1.4 < h \le 1.5$	$1.5 < h \le 1.6$	$1.6 < h \le 1.7$	$1.7 < h \le 1.8$
Frequency	19	25	33	16	12

2 Jason constructs the following frequency table to collect data about the amount of time his classmates spend looking at smart phones or computer screens before school.

Time, t (min)	Tally	Frequency
$0 \le t \le 10$		
$10 \le t \le 20$		
$25 \le t \le 30$		
$30 \le t \le 35$		
$35 \le t \le 50$		

a) Identify three problems with his frequency table.

b) Draw a table which would be suitable for collecting these data.

3 Fran records the number of visitors to a beach on 20 different days.

| 615 | 402 | 552 | 589 | 376 | 412 | 500 | 626 | 436 | 541 |
| 727 | 689 | 543 | 457 | 360 | 476 | 591 | 641 | 425 | 438 |

   a) Design a frequency table with equal width classes for recording the data.

   b) Complete your frequency table.

4 The compound bar chart shows the number and type of books sold by a book shop one week.

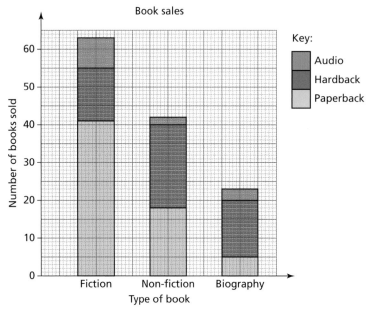

   a) Write down the number of non-fiction books sold.

   b) Find the number of hardback biographies sold.

   c) Find how many more paperback fiction books were sold than hardback fiction books.

   d) Find the total number of paperback books sold last week.

   e) Explain why a compound bar chart is a suitable form of representation of the data.

5 The diagram shows the speeds of all the cars passing a school between 8 a.m. and 9 a.m. one morning.

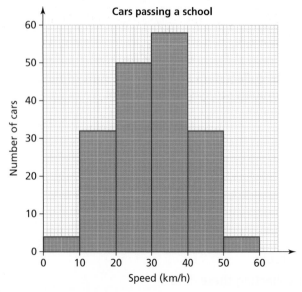

a) The headteacher wants to know if any cars passed the school travelling at more than 50 km/h. Make a conclusion from the graph about this.

b) The speed limit outside the school is 40 km/h. A parent wants to know if more than 25% of cars went faster than 40 km/h. Make a conclusion from the graph about this.

6 A teacher gives her class a test at the start and end of a term. She thinks that her students will do better on the end of term test than on the test at the start of term. The stem-and-leaf diagram shows the marks obtained by the class on the test at the start of term.

Stem	Leaf
5	1 3 5 6
6	4 7
7	3
8	1 5
9	2

Key: 5 │ 1 = 51%

The mean mark in the end of term test is 65%

Was the teacher correct to think that the students will do better on the end of term test? Show clearly how you reached your answer.

7 The stem-and-leaf diagram shows the number of passengers on each of 30 buses.

a) What was the smallest number of passengers on these buses?

b) Draw a frequency diagram to show these data.

c) What advantage does a stem-and-leaf diagram have over a frequency diagram?

Stem	Leaf
0	6 7 9
1	1 1 2 5 6 7 8
2	0 2 3 4 5 6 6 7 9
3	1 1 2 6 7 8
4	0 1 4 5 6

Key: 3 │ 1 = 31 passengers

End of chapter reflection

You should know that...	You should be able to...	Such as...
Continuous data can be summarised in tables with class intervals defined using inequalities. Grouped continuous data can be presented in a frequency diagram.	Summarise continuous data in frequency tables. Draw frequency diagrams to illustrate continuous data presented in a frequency table.	Design and complete a frequency table suitable for recording these ages (in years). 17, 23, 41, 52, 38, 30, 53, 36, 29, 40, 32, 37, 49, 54, 27, 35, 43, 38, 24, 16, 39, 44, 53, 33, 35 Draw a frequency diagram to represent the data.

Dual bar charts and compound bar charts can be used to show information from a two-way table.	Draw and interpret dual bar charts. Draw and interpret compound bar charts with 2 or 3 stacked categories.	Draw a dual bar chart and a compound bar chart to show the information in the table.
		<table><tr><td></td><td>Apple</td><td>Banana</td><td>Grapes</td></tr><tr><td>Men</td><td>27</td><td>54</td><td>41</td></tr><tr><td>Women</td><td>45</td><td>38</td><td>38</td></tr></table>
A stem-and-leaf diagram is a way of presenting data that retains all the actual data values.	Draw and interpret a stem-and-leaf diagram.	The stem-and-leaf diagram shows the lengths (in minutes) of 10 films. Stem | Leaf 9 | 3 10 | 0 5 6 8 11 | 5 9 12 | 1 3 5 **Key:** 9 | 3 = 93 minutes How many of the films are over 2 hours in length?

9 Functions and formulae

You will learn how to:

- Understand that a function is a relationship where each input has a single output. Generate outputs from a given function and identify inputs from a given output by considering inverse operations (including fractions).
- Understand that a situation can be represented either in words or as a formula (mixed operations) and manipulate using knowledge of inverse operations to change the subject of a formula.

Starting point

- what a function is?

    For example, the function 'add 3 to a number' shows for every input value, the output value is 3 more than the input value. A function can be written as:

    a function machine: input → ⟦ add 3 ⟧ → output

    a mapping: $x \mapsto x + 3$

    a formula: $y = x + 3$

- how to generate outputs from a given function and represent the results in a table?

    For example, the function $x \mapsto x + 5$ has the following input-output table:

Input	Output
1	6
2	7
6	11
10	15

- how to identify inputs from a given output by considering inverse operations?

    For example, for the function $x \mapsto 3x$, if the output is 21, the input is 7.

- how to substitute positive integers into expressions?

    For example, write an expression for the perimeter of an equilateral triangle with side length x and use your expression to find the perimeter when $x = 6$.

- the correct order of operations for calculating with numbers and algebra?

    For example, calculate $4 + 2 \times 8$

- how to solve an equation?

    For example, solve $3x + 1 = 16$

- you learn to set up equations that show direct proportion
- you use the relationship between two quantities to make predictions
- you use formulae in mathematics and other subjects
- you learn how to substitute any number into expressions and formulae.

9.0 Getting started

- Here is a function machine.

Complete the input-output table for this function machine.

Input	Output
1	
2	
3	
4	

- Design some more function machines with two operations.
  Each function machine should have the form:

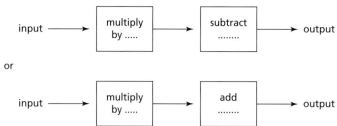

or

Fill in an input-output table (with inputs 1, 2, 3 and 4) for each of your function machines.

- Investigate how the outputs are related to the numbers you used in your function machine.

9.1 Functions

Key terms

A function can be represented in different ways:

As a function machine

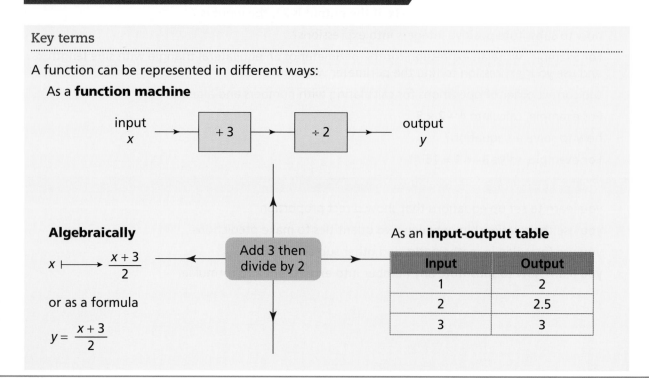

Algebraically

$$x \longmapsto \frac{x+3}{2}$$

Add 3 then divide by 2

or as a formula

$$y = \frac{x+3}{2}$$

As an input-output table

Input	Output
1	2
2	2.5
3	3

As a **mapping diagram**

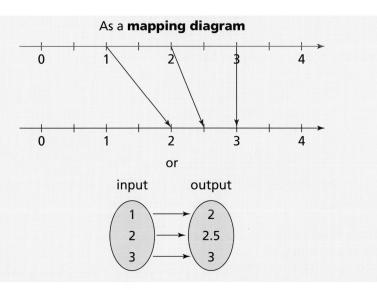

or

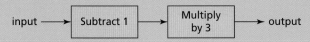

Worked example 1

Here is a function machine.

input → Subtract 1 → Multiply by 3 → output

a) Find the output when the input is 1.25

b) Find the input when the output is −12

a) When the input is 1.25:	The function machine tells you to take the input, subtract 1 and then multiply by 3.	1.25 → Subtract 1 → 0.25 → Multiply by 3 → 0.75
1.25 − 1 = 0.25		
0.25 × 3 = 0.75		
So, the output will be 0.75		
b) When the output is −12:	You need to work backwards to find the input when the output is −12.	? → Subtract 1 → Multiply by 3 → −12
− 12 ÷ 3 = − 4		
− 4 + 1 = − 3	Reverse the function machine by using the inverse operations.	−3 ← Add 1 ← −4 ← Divide by 3 ← −12
So, the input was −3		

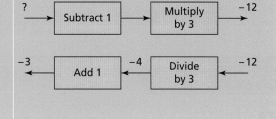

Worked example 2

The input-output table for a function is shown below.

Input	Output
1	−1
2	2
3	5
4	8

a) Describe in words the rule that maps each input value to the corresponding output value.

b) Write a formula to connect the output y with the input value x.

a) The rule is 'multiply by 3 and then subtract 4'.	The difference between the input values is 1 and the difference between the output values is 3. This tells you that the first part of the rule is 'multiply by 3'. To map 4 to 8, the entire rule must be 'multiply by 3 and then subtract 4'.	input values are 1 apart $-3\ -2\ -1\ \ 0\ \ 1\ \ 2\ \ 3\ \ 4\ \ 5\ \ 6\ \ 7\ \ 8$ $-3\ -2\ -1\ \ 0\ \ 1\ \ 2\ \ 3\ \ 4\ \ 5\ \ 6\ \ 7\ \ 8$ output values are 3 apart
b) The function can be written as the formula $y = 3x - 4$.	The rule can be shown as a function machine. When x is fed through the function machine, the output is the expression $3x - 4$. We write this as $x \mapsto 3x - 4$ or $y = 3x - 4$	input x — $\times 3$ — $3x$ — -4 — output $y = 3x - 4$

Exercise 1

1 Here are some functions expressed as function machines.
Express each function algebraically in the form $x \mapsto$

a) $x \rightarrow \boxed{\times 3} \rightarrow \boxed{+5} \rightarrow$

b) $x \rightarrow \boxed{\times 4} \rightarrow \boxed{-2} \rightarrow$

c) $x \rightarrow \boxed{\times 9} \rightarrow \boxed{\div 4} \rightarrow$

d) $x \rightarrow \boxed{\div 3} \rightarrow \boxed{+1} \rightarrow$

e) $x \rightarrow \boxed{+4} \rightarrow \boxed{\times 3} \rightarrow$

f) $x \rightarrow \boxed{-5} \rightarrow \boxed{\div 4} \rightarrow$

2 For each function below:

- draw its function machine
- complete an input-output table, using the inputs 1, 2, 3 and 4.

a) $x \mapsto 5x + 4$

b) $y = 7x - 2$

c) $y = \dfrac{x}{2} + 1$

d) $x \mapsto 4(x - 3)$

e) $y = 5(x + 6)$

f) $x \mapsto \dfrac{3x}{2}$

3 Copy and complete the input-output table for each function machine.

a) input → Add 3 → Divide by 2 → output   b) input → Multiply by 10 → Subtract 7 → output

Input	Output
1	2
11	7
0.6	
9	
1.4	

Input	Output
1	3
$\frac{1}{2}$	
$\frac{4}{5}$	
−3	
	193

4 a) Find the outputs when 0, 1, 2 and 3 are used as the inputs in the function $x \mapsto 2(x - 1)$.

b) Show your answers to part a) on a mapping diagram.

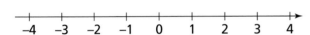

> **Tip**
>
> When the function contains brackets, remember that the operation in brackets is done first.

5 Find the outputs when 1, −4, $\frac{2}{3}$, and 1.2 are used as the inputs in the function $x \mapsto 3x - 1$.

6 Afia is using this function machine:

input → Subtract 1 → Multiply by 3 → output

Afia says, 'I can work out what the input was by multiplying the output by 3 and subtracting 1.'
Do you agree with Afia? Explain your answer.

7 The function $y = 3x - 2$ is represented by the function machine:

input ⟶ ⬡ ⟶ ⬡ ⟶ output

a) Copy and complete the function machine.
b) Find the output when the input is $x = \frac{1}{2}$
c) Find the input when the output is $y = 2.5$

8 A function is represented by the function machine:

input x ⟶ × 10 ⟶ − 9 ⟶ output y

a) Find the output when the input is $x = 0.3$
b) Find the input when the output is $y = 31$
c) Write the relationship between the input x and output y as a formula.

9 Copy and complete the table.

	Function	Formula	Mapping diagram
a)	$x \mapsto \dfrac{x+1}{4}$	$y = \dfrac{x+1}{4}$	15 → , 27 → , 5 →
b)	$x \mapsto \dfrac{x-5}{3}$		14 → , 38 → , 95 →
c)		$y = \dfrac{2x}{5}$	10 → , 35 → , 45 →
d)	$x \mapsto \dfrac{4x-8}{6}$		11 → , 20 → , 3.5 →

10 a) Match each function to the corresponding function machine.

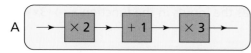

A $\rightarrow$ ×2 $\rightarrow$ +1 $\rightarrow$ ×3 $\rightarrow$

B $\rightarrow$ +1 $\rightarrow$ ×2 $\rightarrow$ ÷3 $\rightarrow$

C $\rightarrow$ ×2 $\rightarrow$ +1 $\rightarrow$ ÷3 $\rightarrow$

D $\rightarrow$ ÷3 $\rightarrow$ +1 $\rightarrow$ ×2 $\rightarrow$

$x \mapsto \dfrac{2x+1}{3}$ W

$x \mapsto \dfrac{2(x+1)}{3}$ X

$x \mapsto 2\left(\dfrac{x}{3}+1\right)$ Y

$x \mapsto 3(2x+1)$ Z

b) For each function, find the output when the input is 3.

11 Find the function that corresponds to each of these mapping diagrams.
Give each answer in the form $x \mapsto$

a)

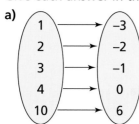

1 → −3
2 → −2
3 → −1
4 → 0
10 → 6

b)

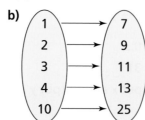

1 → 7
2 → 9
3 → 11
4 → 13
10 → 25

c)

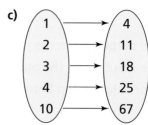

1 → 4
2 → 11
3 → 18
4 → 25
10 → 67

12 Here is a mapping diagram for a function. Find the missing value.

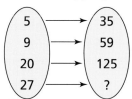

▼ Thinking and working mathematically activity

Jabu has two functions.

Function 1	**Function 2**
$x \mapsto 2x + 6$	$x \mapsto \dfrac{x}{2} + 1$

Jabu inputs a number into Function 1. He then puts the output from Function 1 into Function 2.

Example: Using 5 as the input into Function 1

$$5 \mapsto 2 \times 5 + 6 = 16$$

Use 16 as the input into Function 2

$$16 \mapsto \frac{16}{2} + 1 = 9$$

The output from Function 2 is 9

- Choose your own input values for Function 1 and find what value Jabu would get as the output from Function 2.

- How does the output from Function 2 relate to the input to Function 1?

- Can you explain why the output from Function 2 is related in this way to the input to Function 1?

- Try other pairs of functions, for example,

Function 1	Function 2
$x \mapsto 4x - 8$	$x \mapsto \dfrac{x}{2} + 4$

How does the output from Function 2 relate this time to the input to Function 1?

Discuss

Without doing any calculations, discuss with your partner whether these functions will generate the same output for a given input. Now use examples to check your answer.

input → [+ c] → [− d] → output   input → [× a] → [+ b] → output

and

input → [− d] → [+ c] → output   input → [+ b] → [× a] → output

9.2 Constructing and using formulae

Key terms

A **variable** is a letter that represents an unknown number or value.

A **formula** is a mathematical relationship between two or more variables expressed algrebraically.

A formula does not mean anything unless you say what your variables represent.

Worked example 3

a) Write down a formula to calculate the perimeter *P* of this shape.

b) Use your formula to calculate the perimeter of the shape when *x* = 4 cm and *y* = 2 cm.

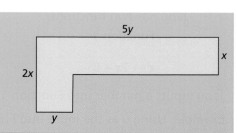

a) $P = 2x + y + x + 4y$ $+ x + 5y$ $P = 4x + 10y$	First calculate the missing lengths in terms of *x* and *y*. Then add together the lengths of each side of the shape to find the perimeter. Write your answer as simply as possible.	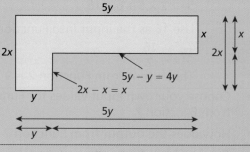
b) $P = 4x + 10y$ $P = 4 \times 4 + 10 \times 2$ $P = 16 + 20$ $P = 36\,cm$	Replace the *x* in the formula with 4 and the *y* in the formula with 2 to find the perimeter. Remember to include units when needed.	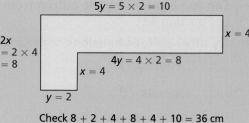 Check $8 + 2 + 4 + 8 + 4 + 10 = 36$ cm

Think about

Can you think of any formulae that you have seen or met before? Write them down.

Exercise 2

1 a) Write a formula for the perimeter *P* of a rectangle with width *w* and length *l*.

b) Use your formula to calculate the perimeter of a rectangle with width 8 cm and length 6 cm.

2 a) Write a formula for the perimeter *P* of this shape.

b) Use your formula to calculate the perimeter when *a* = 3 and *b* = 2.

Discuss

How else can you write the formula for the perimeter of the rectangle?

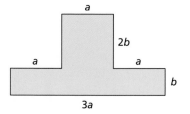

3 **a)** Write a formula for the perimeter P of this shape.

b) Use your formula to calculate the perimeter when $c = 5$ and $d = 4$.

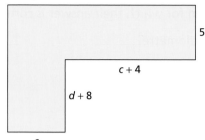

4 To convert a temperature given in degrees Celsius (°C) to degrees Fahrenheit (°F) you multiply the temperature in Celsius by 1.8 and then add 32.

a) Which of these formulae can be used to convert temperatures given in degrees Celsius to degrees Fahrenheit?

$$F = 1.8 + 32C \qquad F = \frac{9}{5}C + 32 \qquad F = \frac{9C + 32}{5}$$

$$F = 1.8C + 32 \qquad F = 1.8(C + 32) \qquad F = 32 + 1.8C$$

b) Use your formula to convert the following temperatures from °C to °F:

   i) 0 °C   **ii)** 20 °C   **iii)** 25 °C   **iv)** 33 °C

c) Oven temperatures can sometimes be given as a gas mark, sometimes in degrees Celsius and sometimes in degrees Fahrenheit.

Here is a table showing some oven temperatures. Use your formula from part **a)** to help you complete a copy of the table.

Gas Mark	Fahrenheit (°F)	Celsius (°C)
1	275	
2		150
3	325	
4		190

> **Did you know?**
>
> Fahrenheit is now only used in the USA, Belize and Jamaica. Celsius is the standard for the rest of the world. However, in certain scientific fields such as astronomy, the Kelvin scale is used. Zero Kelvin is equivalent to −273.15 °C (or −459.67 °F).

5 In this rectangle the length is three times the width.

w is the width of the rectangle.

Bert says, 'The perimeter of the rectangle $= w + 3w + w + 3w = 8w$.'

Claire says, 'The perimeter of the rectangle $= w + w + 3 + w + w + 3 = 4w + 6$.'

a) Who do you agree with? Explain your answer.

b) For the person who has made a mistake, write a question for which their answer is correct.

6 A taxi driver charges a fixed fee of $3.50 plus 20 cents per kilometre.

a) Write a formula for the total cost C for a journey k kilometres long.

b) Use your formula to find the cost of a 12 kilometre journey.

c) A recent journey cost $8.50. Find how far the taxi travelled.

7 A phone company charges $15 per month, 5 cents for each text message and 2 cents per minute for phone calls.

a) Write a formula for the total bill (b), where t texts are sent and c minutes of calls are made.

b) Use your formula to find the cost in a month where 25 text messages were sent and 10 minutes of phone calls were made.

c) In one month, Denise's total bill was $18 and she had 30 minutes of phone calls. How many text messages much she have sent?

8 The sum of the angles in a polygon with n sides is given by the formula:
sum of angles $= (n - 2) \times 180°$

a) Find the sum of the angles in a heptagon (7 sides).

b) What is the name of the polygon whose angle sum is 540°?

9 The cost $c of buying b loaves of bread and having them delivered is given by the formula:
$c = 2b + 5$

A cafe uses 100 loaves of bread each week. They can buy 50 loaves twice a week or buy all 100 loaves once a week. Calculate the difference in cost for the week.

Thinking and working mathematically activity

The sum of the numbers that the shape covers is
$27 + 34 + 35 + 36 = 132$

- Move the shape to different positions on the grid. Always keep the shape the same way up.
  Find the sum of the numbers the shape covers each time.
  Try to make a conclusion about the sums you can get. Is it, for example, possible for the sum of the numbers to be 57?

- Use n to represent the smallest number covered.
  Find a formula in terms of n for the sum, S, of the numbers covered.

1	2	3	4	5	6	7	8
9	10	11	12	13	14	15	16
17	18	19	20	21	22	23	24
25	26		28	29	30	31	32
33			37	38	39	40
41	42	43	44	45	46	47	48
49	50	51	52	53	54	55	56
57	58	59	60	61	62	63	64
65	66	67	68	69	70	71	72
73	74	75	76	77	78	79	80

- Now investigate rotating the shape. For example, consider the arrangement shown below.

  Find a new formula for the sum, S, of the numbers covered. Each time, write your formula in terms of the smallest number covered, n.

1	2	3	4	5	6	7	8
9	10	11	12	13	14	15	16
17	18	19	20	21	22	23	24
25	26	27	28		30	31	32
33	34	35			38	39	40
41	42	43	44		46	47	48
49	50	51	52	53	54	55	56
57	58	59	60	61	62	63	64
65	66	67	68	69	70	71	72
73	74	75	76	77	78	79	80

9.3 Changing the subject

Key terms

Here are some formulae.

$$y = 3x + 2 \qquad y = ax - 9 \qquad y = \frac{2x + 7}{9}$$

y is called the **subject** of each of these formulae. The subject is the variable that appears on its own and is written equal to everything else. The subject is usually written on the left of the equals sign.

Worked example 4

Make x the subject of the formula $A = 38x - 24$

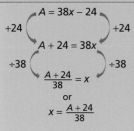

You need to reverse the operations. First add 24 to both sides of the formula.

Now divide both sides by 38.

You can reverse the order now so that you have the formula written in the format required.

'Make x the subject' means to rearrange the formula so that x is written in terms of A.

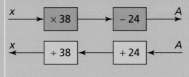

1. Make x the subject of each formula.

   a) $y = x + 4$ b) $y = 5x$ c) $y = x - a$ d) $y = \frac{x}{3}$

   e) $y = 7x$ f) $y = \frac{x}{5}$ g) $y = x + 2b$ h) $y = x - 3c$

   i) $y = 3x + 5$ j) $y = 2x - a$ k) $3y = 2x - 5b$ l) $y = \frac{x + 7}{3}$

   m) $y = \frac{x - 4}{5}$ n) $y = \frac{a + x}{4}$ o) $y = \frac{x}{5} - a$ p) $y = \frac{x}{2} + 3b$

2. Match the formula with the correct rearranged one.

   a) $y = 3x - 1$ a) $x = 5(y - 3)$

   b) $y = 4(x + 3)$ b) $x = \frac{y + 1}{3}$

   c) $y = 2x + 3$ c) $x = \frac{y}{4} - 3$

   d) $y = \frac{x}{5} + 3$ d) $x = \frac{y - 3}{2} - 3$

3. Make x the subject of each formula.

   a) $y = 4x + 3$ b) $y = 3(x - 2)$ c) $y = ax - 4$ d) $y = \frac{x - 2}{5}$

4. $y = mx + c$ is the formula for the equation of a straight line. Make x the subject of this formula.

5. Anika and Jasmine are both trying to make q the subject of this formula: $p = \frac{q}{3} - 2$

   Both girls say that you first add 2 to both sides, giving $p + 2 = \frac{q}{3}$

   Both girls say that you then multiply by 3

   Anika says the new formula is $q = 3p + 2$

   Jasmine says it is $q = 3(p + 2)$

   Who is correct, Anika or Jasmine? Give a reason for your answer.

Thinking and working mathematically activity

Compare these two methods for making r the subject.

Method 1: Expand the brackets first

$t = p(r - 2)$

$t = pr - 2p$

$t + 2p = pr$

$r = \frac{t + 2p}{p}$

Method 2: Divide by the coefficient in front of the brackets

$t = p(r - 2)$

$\frac{t}{p} = r - 2$

$r = \frac{t}{p} + 2$

Try the methods on these formulae to make *r* the subject:

- $t = 4(r - 3)$
- $t = 3(2r + 5)$
- $t = 5(ar - 7)$

Which method do you prefer?

What are the advantages and disadvantages of each method?

Consolidation exercise

1 The input *x* and output *y* of a function are connected by the formula

$y = \dfrac{x}{4} - 3$

Copy and complete the mapping diagram.

input, *x*   output, *y*

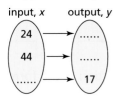

2 Copy and complete the statements so that each function is expresssed algebraically.

a)
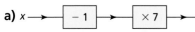

$x \mapsto ....(x ....)$

b)

Input	Output
1	9
2	14
3	19

$x \mapsto 5x ....$

c)

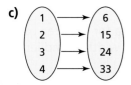

$x \mapsto ....$

3 A function machine is

$x \longrightarrow \boxed{-1} \longrightarrow \boxed{\times 4} \longrightarrow$

Find the output when then input is

a) –6   **b)** $3\frac{1}{2}$ **c)** 1.25

4 Make x the subject of these formulae:

a) $y = x - 7$ **b)** $y = \dfrac{x}{4} + 5$ **c)** $y = 2x + 3$ **d)** $y = \dfrac{x + 1}{3}$

5 A function is defined as $x \mapsto 11x - 120$.

Find the value of the input which has the following property:
when the input is *x*, the output is also *x*.

6 There are two different delivery companies who price their service as follows:

Deliverme   Fee of $5 plus 25 cents per mile

PackagesRUs   Fee of $2 plus 40 cents per mile

Write a formula for each company for the price of delivery.

7 Tom rearranges the formula $R = 2C + 14$ to make C the subject. Which one is incorrect? Give a reason for your choice.

$$C = \frac{R - 14}{2} \qquad C = \frac{R}{2} - 7 \qquad C = 2(R - 14) \qquad C = \frac{1}{2}(R - 14)$$

End of chapter reflection

You should know that...	You should be able to...	Such as...
Functions can be specified in several ways, including as function machines and algebraically.	Find an algebraic rule that matches a function machine. Find the input when given the output of a function machine.	A function is defined by the function machine: $\longrightarrow \boxed{+7} \longrightarrow \boxed{\div 5} \longrightarrow$ a) Write this function in the form $x \mapsto$ .... b) Find the input when the output is 1.
	Find a formula that matches a mapping diagram.	For the mapping diagram, write the relationship between the inputs (x) and outputs (y) as a formula. 1 → −1 2 → 5 3 → 11 4 → 17
A variable is a letter that represents a number that can take different values.	Write a formula to connect two variables.	Write a formula for the area of a rectangle with width w and length l.
A formula represents a mathematical relationship between two or more variables.	Use a simple formula to find missing information.	The area A of a triangle with base b and height h is found using the formula $A = \dfrac{bh}{2}$ Find the area of a triangle with base 7 cm and height 4 cm.
The subject of a formula is the variable that appears on its own and is made equal to all other things in the formula.	Change the subject of a formula.	Make a the subject of the formula $m = 4a - 11$

10

Fractions

You will learn how to:

- Estimate and subtract mixed numbers and write the answer as a mixed number in its simplest form.
- Estimate and multiply an integer by a mixed number, and divide an integer by a proper fraction.

Starting point

Do you remember…

- how to estimate and add mixed numbers?

  For example, estimate and then calculate $3\frac{3}{5} + 2\frac{4}{9}$, writing the answer as a mixed number in its simplest form.

- how to estimate, multiply and divide proper fractions?

  For example, estimate and then calculate $\frac{2}{3} \times \frac{7}{8}$ and $\frac{5}{6} \div \frac{3}{4}$

This will also be helpful when…

- you learn to add and subtract proper and improper fractions, and mixed numbers, using the order of operations
- you learn to multiply and divide mixed numbers.

10.0 Getting started

Here are six chocolate bars.

Find how many people can share the chocolate if each person gets (a) half a bar (b) one third of a bar.

Describe a method for dividing a whole number by a proper fraction. Use your method to find (a) the number of half bars in 10 chocolate bars (b) the number of quarter bars in 8 chocolate bars.

Here are $1\frac{2}{3}$ chocolate bars.

Find how many bars there are altogether if (a) 3 people each have $1\frac{2}{3}$ bars (b) 4 people each have $1\frac{2}{3}$ bars.

Describe a method for multiplying a mixed number by a whole number. Use your method to find (a) the total amount if 3 people each have $1\frac{1}{2}$ chocolate bars (b) the total amount if 5 people each have $2\frac{1}{4}$ chocolate bars.

10.1 Subtracting mixed numbers

Worked example 1

Estimate and then calculate these subtractions.

Write each answer as a mixed number with the fraction in its simplest form.

a) $3\frac{2}{3} - 2\frac{1}{5}$

b) $3\frac{1}{4} - 1\frac{5}{8}$

a) Estimate: $3\frac{2}{3} - 2\frac{1}{5} \approx 4 - 2 = 2$	To estimate the answer, round each mixed number to the nearest whole number. $3\frac{2}{3}$ is closer to 4 than to 3 $2\frac{1}{5}$ is closer to 2 than to 3	
$3\frac{2}{3} - 2\frac{1}{5}$	3 is greater than 2 $\frac{2}{3}$ is greater than $\frac{1}{5}$	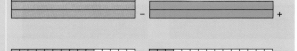
$3 - 2 = 1$ $\frac{2}{3} - \frac{1}{5} = \frac{10}{15} - \frac{3}{15} = \frac{7}{15}$ $3\frac{2}{3} - 2\frac{1}{5} = 1\frac{7}{15}$	So, you can separately find $3 - 2$ and $\frac{2}{3} - \frac{1}{5}$ Add the two results to find the answer.	
b) $3\frac{1}{4} - 1\frac{5}{8} \approx 3 - 2 = 1$ $3\frac{1}{4} - 1\frac{5}{8}$	$3\frac{1}{4}$ is closer to 3 than to 4 $1\frac{5}{8}$ is closer to 2 than to 1 3 is greater than 1, but $\frac{1}{4}$ is less than $\frac{5}{8}$	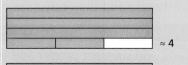 

		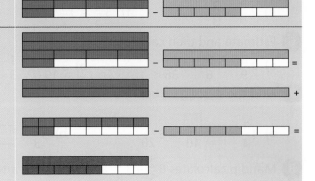
$3\frac{1}{4} - 1\frac{5}{8} = 2\frac{5}{4} - 1\frac{5}{8}$	Write one of the 1s of $3\frac{1}{4}$ as quarters: $3\frac{1}{4} = 2 + 1\frac{1}{4} = 2 + \frac{5}{4}$ $\frac{5}{4}$ is greater than $\frac{5}{8}$	
$2 - 1 = 1$ $\frac{5}{4} - \frac{5}{8} = \frac{10}{8} - \frac{5}{8} = \frac{5}{8}$ $3\frac{1}{4} - 1\frac{5}{8} = 1\frac{5}{8}$	Separately find $2 - 1$ and $\frac{5}{4} - \frac{5}{8}$ Add the two results to find the answer.	

Discuss

An alternative method for both parts of Worked example 1 is to write both mixed numbers as improper fractions, do the subtraction and then convert the answer to a mixed number. Try this method yourself.

Did you know?

The commutative law tells you that changing the order of numbers in an addition or multiplication does not change the result.

Exercise 1

 1–3, 5–9

1 Jasper says that $\frac{12}{5}$ is equivalent to $1\frac{2}{5}$. Do you agree with Jasper? Explain your answer.

2 Which of these improper fractions is equivalent to $4\frac{3}{5}$?

    A: $\frac{23}{20}$ **B:** $\frac{12}{5}$ **C:** $\frac{23}{5}$ **D:** $\frac{12}{20}$ **E:** $\frac{15}{512}$

3 Do these subtractions.

If an answer is greater than 1, write it as a mixed number. Write all fractions in their simplest form.

    a) $3\frac{4}{5} - 1\frac{1}{5}$ **b)** $1\frac{1}{2} - \frac{3}{4}$ **c)** $2\frac{1}{2} - 1\frac{1}{4}$ **d)** $4\frac{3}{8} - 1\frac{7}{8}$

4 Estimate the answer to each calculation below.

Then use a calculator to do the calculation.

    a) $2\frac{8}{9} - 1\frac{4}{5}$ **b)** $7\frac{1}{3} - 2\frac{5}{8}$ **c)** $4\frac{4}{7} - \frac{3}{5}$ **d)** $5\frac{3}{11} - 3\frac{3}{4}$

5 Estimate and then calculate each answer.

If an answer is greater than 1, write it as a mixed number. Write all fractions in their simplest form.

a) $5\frac{1}{2} - 2\frac{1}{3}$ b) $2\frac{8}{15} - 1\frac{4}{5}$ c) $5\frac{1}{15} - 1\frac{2}{3}$ d) $2\frac{1}{2} - 1\frac{3}{8}$

e) $2\frac{4}{9} - \frac{2}{3}$ f) $5\frac{4}{5} - 3\frac{1}{2}$ g) $4\frac{7}{8} - 2\frac{2}{3}$ h) $6\frac{1}{7} - 2\frac{4}{5}$

6 Estimate and then calculate:

a) $2\frac{8}{9} - 1\frac{5}{6} + 1\frac{13}{36}$ b) $3\frac{9}{10} - 1\frac{1}{3} - 1\frac{1}{5}$

7 Find the missing values.

a) $1\frac{3}{4} + \boxed{}\frac{\boxed{}}{10} = 5\frac{1}{20}$ b) $3\frac{2}{3} + \boxed{}\frac{\boxed{}}{2} + 3\frac{1}{12} = 8\frac{1}{4}$

> **Tip**
>
> If $3 + \square = 5$, what would the $\square$ be?

8 Mandip calculates $6\frac{2}{5} - 3\frac{7}{8}$ and writes the answer as a mixed number.

Her working is below and her answer is correct.

$$6\frac{2}{5} - 3\frac{7}{8} = \frac{32}{5} - \frac{31}{8}$$

$$= \frac{256}{40} - \frac{155}{40}$$

$$= \frac{101}{40}$$

$$= 2\frac{21}{40}$$

a) Critique her method, describing any disadvantages.

b) Show a more efficient method for calculating $6\frac{2}{5} - 3\frac{7}{8}$.

Thinking and working mathematically activity

Manon and Roald want to find $3\frac{1}{10} - 2\frac{4}{5}$

Roald says, 'To find the answer, I calculate $3 - 2 = 1$ and $\frac{1}{10} - \frac{4}{5} = \frac{1}{10} - \frac{8}{10} = -\frac{7}{10}$.
Then I add the two results.'

Manon says, 'To find the answer, I calculate $\frac{1}{5} + \frac{1}{10}$.'

Investigate these two methods. For each method, say whether it works and explain why.

Evaluate the two methods. Compare them with the other methods that you know.

Discuss whether you would choose different methods for different subtraction calculations.

Key terms

To **partition** a number is to break it into parts that are added or subtracted.
For example, you can partition 7 into 2 + 5, 1 + 6, 10 − 3, and so on.

The **distributive** law tells you that to multiply by a number, you can partition the number first and then multiply by each part. For example, 6 × 13 = 6 × 10 + 6 × 3 and 6 × 19 = 6 × 20 − 6 × 1.

To **invert** a fraction is to swap the numerator and denominator.
For example, if you invert $\frac{5}{8}$ you get $\frac{8}{5}$.

The **reciprocal** of a number equals 1 divided by the number.
The product of a number and its reciprocal is 1.

For example, the reciprocal of 2 is $1 \div 2 = \frac{1}{2}$. (Notice that $2 \times \frac{1}{2} = 1$). You can find the reciprocal of a fraction by inverting it. For example, the reciprocal of $\frac{2}{3}$ is $\frac{3}{2}$. (Notice that $\frac{2}{3} \times \frac{3}{2} = 1$).

Worked example 2

Estimate and then calculate $3 \times 2\frac{3}{4}$

$3 \times 2\frac{3}{4} \approx 3 \times 3 = 9$	Round the fraction to the nearest whole number.	≈ 3
$3 \times 2\frac{3}{4} = (3 \times 2) + (3 \times \frac{3}{4})$	Use partitioning and the distributive law to split the calculation into two parts.	$3 \times$
$3 \times 2 = 6$ $3 \times \frac{3}{4} = \frac{9}{4} = 2\frac{1}{4}$	Calculate each part.	
$6 + 2\frac{1}{4} = 8\frac{1}{4}$	Add the parts to find the answer.	

Worked example 3

a) $2 \div \frac{2}{5}$

b) $2 \div \frac{3}{5}$

---

a)

$2 \div \frac{2}{5} = 2 \times \frac{5}{2}$

$\qquad = \frac{10}{2}$

$\qquad = 5$

| Dividing by $\frac{2}{5}$ is the same as multiplying by $\frac{5}{2}$. |

Each of the ones can be divided into five fifths:

$2 = \frac{10}{5}$

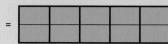

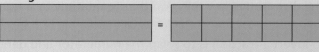

$\frac{10}{5} \div \frac{2}{5} = 5$

There are five $\frac{2}{5}$s in 2.

---

b)

$2 \div \frac{3}{5} = 2 \times \frac{5}{3}$

$\qquad = \frac{10}{3}$

$\qquad = 3\frac{1}{3}$

| Dividing by $\frac{3}{5}$ is the same as multiplying by $\frac{5}{3}$. There is not a whole number of $\frac{3}{5}$s in 2. |

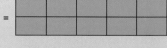

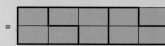

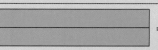

$\frac{10}{5} \div \frac{3}{5} = 10 \div 3$

$= \frac{10}{3}$ or $3\frac{1}{3}$

There are $3\frac{1}{3}$ lots of $\frac{3}{5}$ in 2.

Exercise 2 1–8

1 Estimate and then calculate each answer.

Write your answers as mixed numbers, with the fractions in their simplest form.

a) $2 \times 1\frac{1}{3}$ **b)** $4 \times 1\frac{1}{4}$ **c)** $1\frac{3}{4} \times 4$

d) $10 \times 3\frac{1}{5}$ **e)** $5 \times 2\frac{2}{3}$ **f)** $2\frac{5}{6} \times 3$

2 Estimate and then calculate each answer.

Write your answers as improper fractions in their simplest form.

a) $2\frac{1}{6} \times 4$ **b)** $3 \times 1\frac{1}{2}$ **c)** $2 \times 2\frac{3}{5}$

d) $2\frac{7}{10} \times 5$ **e)** $2 \times 1\frac{5}{8}$ **f)** $1\frac{1}{6} \times 10$

3 Emilie finds the value of $11\frac{2}{7} \times 3$ as a mixed number.

$$11\frac{2}{7} \times 3 = \frac{79}{7} \times 3$$

$$= \frac{237}{7}$$

$$= 33\frac{6}{7}$$

Write a more efficient method.

4 Ricardo writes the following:

$$1\frac{1}{3} \times 5 = 1\frac{5}{3}$$

a) Describe the two mistakes Ricardo has made.

b) Show the correct working.

5 Complete each statement using <, = or >.

a) $5 \times 2\frac{5}{6} \square 4 \times 6\frac{1}{8}$

b) $3 \times 2\frac{1}{2} \square 2 \times 3\frac{1}{2}$

c) $2\frac{1}{6} \times 4 \square 4 \times 2\frac{1}{2}$

d) $2 \times 1\frac{4}{5} \square 4 \times 1\frac{1}{3}$

e) $4\frac{1}{3} \times 3 \square 3\frac{1}{3} \times 4$

f) $4\frac{5}{7} \times 5 \square 5 \times 4\frac{5}{7}$

> **Tip**
>
> You do not always need to find the exact answer to each calculation.

> **Think about**
>
> If you multiply an integer by a mixed number, is the answer:
> - always, sometimes or never greater than the integer
> - always, sometimes or never greater than the mixed number
> - always, sometimes or never greater than 1?

6 Find:

a) $8 \div \frac{1}{2}$

b) $4 \div \frac{1}{3}$

c) $2 \div \frac{1}{4}$

d) $5 \div \frac{1}{5}$

e) $6 \div \frac{1}{4}$

f) $6 \div \frac{3}{4}$

g) $4 \div \frac{1}{5}$

h) $4 \div \frac{2}{5}$

> **Think about**
>
> If you divide an integer by a proper fraction, is the answer bigger or smaller than the original integer? Explain why.

7 Calculate these divisions.

If the answer is not a whole number, write it as a mixed number.

a) $8 \div \frac{2}{3}$

b) $7 \div \frac{7}{8}$

c) $6 \div \frac{3}{4}$

d) $4 \div \frac{4}{7}$

e) $3 \div \frac{2}{3}$

f) $2 \div \frac{3}{4}$

g) $2 \div \frac{3}{5}$

h) $3 \div \frac{4}{5}$

8 Explain why each answer in question 7 is greater than the integer.

9 Sami divides an integer by a fraction.

The answer is 10.

$$\square \div \dfrac{\square}{\square} = 10$$

Write down what the integer and the fraction could be. Now think of a different pair of answers.

10 Technology question Estimate the answer to each calculation below.

Then use a calculator to do the calculation. Write answers greater than 1 as mixed numbers.

a) $3 \times 2\dfrac{5}{8}$

b) $4\dfrac{6}{7} \times 4$

c) $2\dfrac{1}{9} \times 6$

d) $5 \times 3\dfrac{4}{5}$

e) $18 \div \dfrac{2}{3}$

f) $21 \div \dfrac{7}{12}$

g) $16 \div \dfrac{4}{5}$

h) $8 \div \dfrac{8}{11}$

Thinking and working mathematically activity

Use the fact that $2 \div \dfrac{1}{3} = 6$ to find:

$2 \div \dfrac{2}{3}$ $\qquad$ $4 \div \dfrac{1}{3}$ $\qquad$ $\dfrac{2}{5} \div \dfrac{1}{3}$

Explain your reasoning.

Consolidation exercise

1 Is the following statement always, sometimes or never true? Explain your answer.

Fractions with larger denominators have smaller values.

2 For a school's sports day, a group of students made $12\dfrac{1}{2}$ litres of lemonade.
At the end of the day they had $2\dfrac{5}{8}$ litres left over.

Work out how many litres of lemonade they gave out.

3 Use the digits 2, 3, 4, 5, 6 and 7 each exactly once to make the following calculation true.

 $= 5\dfrac{1}{12}$

4 Brogan has 4 tins of paint. Each tin contains $1\dfrac{1}{3}$ litres of paint. Find how many litres of paint she has. Write your answer as a mixed number.

5 Find the missing numbers in the calculations.

a) $2\dfrac{1}{6} \times \square = \dfrac{39}{6}$

b) $1\dfrac{\square}{5} \times 6 = 9\dfrac{3}{5}$

c) $\dfrac{\square}{3} \times 9 = 24$

6 Work out the value of each calculation. Put a ring around the odd one in each set.

a) $6 \div \dfrac{1}{2}$ $\qquad$ $3 \div \dfrac{1}{3}$ $\qquad$ $2 \div \dfrac{1}{6}$

b) $20 \div \dfrac{1}{4}$ $\qquad$ $12 \div \dfrac{1}{5}$ $\qquad$ $6 \div \dfrac{1}{10}$

7 Makoto uses $\frac{2}{5}$ kg of flour to make one cake. Find how many cakes he can make using 4 kg of flour.

8 Match each calculation with its answer.

$6 \div \frac{1}{3}$	$2\frac{2}{3}$
$3 \div \frac{2}{7}$	$6\frac{2}{3}$
$4 \div \frac{2}{5}$	10
$2 \div \frac{3}{4}$	$10\frac{1}{2}$
$6 \div \frac{9}{10}$	18

End of chapter reflection

You should know that...	You should be able to...	Such as...
To estimate the result of a mixed number subtraction, round each number to the nearest integer.	Estimate the difference between two mixed numbers.	Estimate $4\frac{1}{5} - 2\frac{3}{4}$
To subtract a mixed number from a second mixed number, first check whether the fraction part of the first number is bigger than the fraction part of the second number. • If it is: subtract the integer parts and the fraction parts separately and add the results. • If it is not: write one of the 1s of the first number as part of the fraction. Then follow the method above.	Subtract one mixed number from another.	Find: **a)** $2\frac{3}{4} - 1\frac{1}{4}$ **b)** $3\frac{2}{5} - 1\frac{7}{10}$
To multiply a mixed number by an integer, either: • multiply the whole number part of the mixed number by the integer, multiply the fraction part by the integer, and add the results, or • write the mixed number as an improper fraction and then multiply.	Multiply a mixed number by an integer.	Find: **a)** $3 \times 2\frac{1}{3}$ **b)** $3\frac{3}{4} \times 5$
To divide an integer by a proper fraction, invert the fraction and multiply.	Divide an integer by a proper fraction.	Find: **a)** $4 \div \frac{2}{3}$ **b)** $5 \div \frac{3}{5}$

11 Length, area and volume

You will learn how to:

- Know that distances can be measured in miles or kilometres, and that a kilometre is approximately $\frac{5}{8}$ of a mile or a mile is 1.6 kilometres.
- Use knowledge of rectangles, squares and triangles to derive the formulae for the area of parallelograms and trapezia. Use the formulae to calculate the area of parallelograms and trapezia.
- Use knowledge of area and volume to derive the formula for the volume of a triangular prism. Use the formula to calculate the volume of triangular prisms.
- Use knowledge of area, and properties of cubes, cuboids, triangular prisms and pyramids to calculate their surface area.

Starting point

Do you remember…

- how to convert between kilometres (km), metres (m), centimetres (cm) and millimetres (mm)?

  For example, convert 3.4 m to cm. Convert 0.14 km to m.

- how to find the area of shapes made from rectangles and simple triangles? For example, find the area of this shape.

  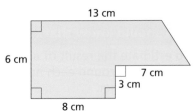

- how to find the volume of a cuboid?
  For example, find the volume of this cuboid.

  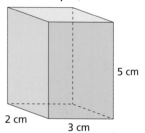

- how to find the surface area of a cuboid by considering its net?
  For example, draw a net of the cuboid above to help you calculate its total surface area.

This will also be helpful when…

- you find the volume and surface area of prisms and more complex 3D shapes.

11.0 Getting started

Before there were accurate measuring instruments, measurements were often based on parts of the body – usually parts of the arm and hands.

One of the oldest measurements is the cubit – the distance between the elbow and the tip of the middle finger in an adult male. It was used in ancient Egypt almost 5000 years ago in the building of the pyramids.

The cubit was divided into seven palms – the width across the palm below the four fingers – and each palm was then divided into four fingers.

Measuring sticks at the time of Tutankhamun would consist of a cubit divided into seven palms and 28 fingers.

- In groups, find things to measure using your own arms and hands for cubits, palms and fingers. Choose one large object such as a wall, one medium-sized object such as table and one small item such as a book.
- Can you think of other ways you could measure distances on the floor? Use these to compare two rooms. Which room is longer? Compare sides of a yard or field to see which is the longest side.

11.1 Converting between miles and kilometres

Key terms

A **mile** is an **imperial unit** of length used to measure large distances.
A mile is approximately equal to 1.6 kilometres. A kilometre is approximately $\frac{5}{8}$ of a mile.

Approximately equal to means almost equal, but not exactly equal to. The symbol ≈ means approximately equal to.

Tip

When converting miles given as a multiple of 5, use the fact that 1 mile is approximately equal to $\frac{8}{5}$ kilometre. Otherwise use the approximation 1 mile ≈ 1.6 kilometres.

Worked example 1

a) Find the approximate value of 72 km in miles.

b) Find the approximate value of 38 miles in km.

c) Find the approximate value of 35 miles in km.

d) Find the approximate value of 14 km in miles.

a) $72 \text{ km} \approx 72 \times \frac{5}{8}$ miles $72 \times \frac{5}{8} = 72 \div 8 \times 5$ $\qquad = 45$ so 72 km ≈ 45 miles	There are approximately 8 km in 5 miles. To convert approximately from km to miles, divide the number of km by 8 and then multiply by 5	$\begin{array}{\|c\|}\hline \text{72 km} \\ \hline \end{array}$ 8 km | 8 km | 8 km | 8 km | 8 km | 8 km | 8 km | 8 km | 8 km 5 miles |5 miles|5 miles|5 miles|5 miles|5 miles|5 miles|5 miles|5 miles 45 miles
b) 38 miles ≈ 38 × 1.6 km 38 × 1.6 = 60.8 so 38 miles ≈ 60.8 km	1 mile is approximately 1.6 km. To convert approximately from miles to km, multiply the number of miles by 1.6	

c) 35 miles ≈ 35 × $\frac{8}{5}$ km

35 ÷ 5 × 8 = 56

so 35 miles ≈ 56 km

The number of miles is a multiple of 5, so use the approximation 1 mile is $\frac{8}{5}$ km.

35 miles						
5 miles	5 miles	5 miles	5 miles	5 miles	5 miles	5 miles
8 km	8 km	8 km	8 km	8 km	8 km	8 km
56 km						

d) 14 km ≈ 14 × 0.625
   = 8.75 miles

1 mile is approximately 1.6 km. 1 km is approximately
1 ÷ 1.6 = 0.625 miles.
To convert from km to miles, multiply the number of km by 0.625 or divide by 1.6

Exercise 1

1–4, 7–11

1 Find the approximate number of km in each of the following.

   a) 10 miles   **b)** 25 miles   **c)** 55 miles   **d)** 250 miles

2 Find the approximate number of miles in each of the following.

   a) 24 km   **b)** 32 km   **c)** 48 km   **d)** 800 km

3 Find the greater distance in each of these pairs.

   a) 240 km or 160 miles   **b)** 400 km or 240 miles   **c)** 46 km or 30 miles

4 A road sign in France says that the distance to Paris is 296 km.
Convert this distance to miles.

5 Calculate the approximate number of miles in each of the following.
Give your answers as decimals.

   a) 26 km   **b)** 40 km   **c)** 130 km   **d)** 250 km   **e)** 1000 km

6 If 1 mile is $\frac{8}{5}$ or 1.6 km, calculate the approximate number of km in each of the following.
Give your answers as decimals.

   a) 36 miles   **b)** 42 miles   **c)** 108 miles   **d)** 273 miles   **e)** 1000 miles

7 Write a distance between 90 km and 100 km that is easy to convert approximately to miles.
Explain what makes the conversion easy.

8 Jamila is checking her sister's work. Without any calculations, she says that all of the following calculations are wrong. Explain how she knows that each calculation is incorrect, and then correct the calculation.

   a) 160 miles ≈ 100 km   **b)** 30 miles ≈ 60 km   **c)** 85 km ≈ 47 miles

9 Explain the mistake that Jamila's sister has made when working out her answer to question 8 a.

10 You are allowed to use your calculator to work out the approximate value of only two of the following distances in miles. You must work out the rest without a calculator. Which two would you choose to use your calculator for? Give a reason for your answer.

a) 80 km   b) 188 km   c) 64.8 km   d) 376 km   e) 602 km

11 Look at the diagram below, which is a rough conversion chart for miles and km.

a) Which side of the line represents kilometres and which represents miles?

b) Use the scale to find the approximate value of each of the following.

   i) 4 km in miles   ii) 3 miles in km   iii) 4.5 miles in km

12 Andrew has enough power in his car to travel 40 miles.
The nearest point he can recharge his car is 68 km away.
Find out if Andrew has enough power in his car to drive
to the recharging point. Show how you found your answer.

▼ Thinking and working mathematically activity

Toby uses the following method for converting 280 km to miles.

> 280 ÷ 2 = 140
>
> 140 ÷ 2 = 70   140 + 35 = 175
>
> 70 ÷ 2 = 35
>
> 280 km = 175 miles

- Explain how Toby's method works.
- Use Toby's method to change the following measurements to miles:

  208 km

  100 km

  68 km

  4.4 km
- Compare Toby's method with other non-calculator methods. Which method do you prefer, and why?

11.2 Area of 2D shapes

Key terms

The **perpendicular height** is the height of a shape measured
at right angles to the base of the shape, as shown in the diagram.

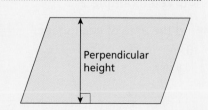

Perpendicular height

Thinking and working mathematically activity

- Draw the parallelogram *PQRS* on squared paper.

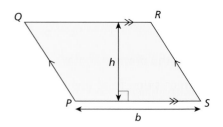

By thinking about the triangle *PQM* at the end of the parallelogram moving to the other end as *SRN*, find a rule for finding the area of a parallelogram.

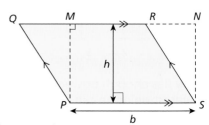

- Draw a triangle on squared paper. Rotate the same triangle and join it to the original as shown to create a parallelogram.
  Use your diagram to find an expression for the area of the triangle in terms of *b* and *h*.

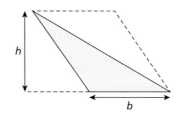

- A trapezium has two parallel sides of *a* and *b*, with a vertical height of *h*.

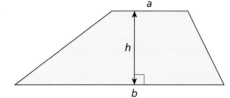

Draw a trapezium on squared paper as shown. Then, by drawing the same trapezium and joining it as shown, create a parallelogram.

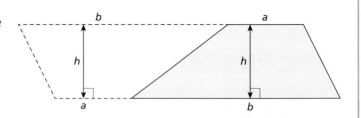

Use your diagram to find an expression for the area of a trapezium in terms of *a*, *b* and *h*.

Worked example 2

Find the area of each shape:

a)
8 cm 7 cm
11 cm

b)
5 m
18 m

c)
8 mm
6 mm
14 mm

---

134   Stage 8: Student's Book

a) Area = base × perpendicular height = 11 × 7 = 77 cm²	The parallelogram can be rearranged into a rectangle as shown. So, the area covered by the parallelogram is the same as that of a rectangle with length 11 cm and width 7 cm. Hence, the area of the parallelogram = 11 × 7 = 77 cm² (We do not need the 8 cm measurement provided.)	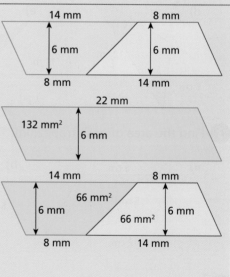
b) Area = $\frac{1}{2}$ base × height = $\frac{1}{2}$ (18 × 5) = $\frac{1}{2}$ (90) = 45 m²	The triangle is exactly half of a parallelogram as shown. The area of the parallelogram is 18 × 5 = 90 m². Since the triangle is half of the parallelogram, the area of the triangle = $\frac{1}{2}$ (90) = 45 m²	
c) Area = $\frac{1}{2}$ (a + b)h = $\frac{1}{2}$ ([14 + 8] × 6) = $\frac{1}{2}$ (22 × 6) = $\frac{1}{2}$ (132) = 66 mm²	The trapezium is exactly half of a parallelogram, as shown. This parallelogram has length 22 mm and perpendicular height 6 mm. Therefore, the area of the parallelogram = 22 × 6 = 132 mm² Since the trapezium is half of the parallelogram, the area of the trapezium = $\frac{1}{2}$ (132) = 66 mm²	

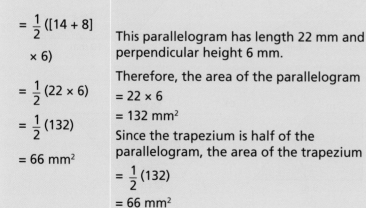

1 Find the area of each parallelogram.

a)

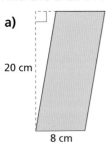

20 cm

8 cm

b)

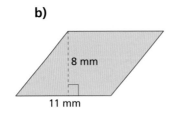

8 mm

11 mm

c)

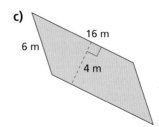

16 m

6 m

4 m

d)

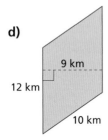

9 km

12 km

10 km

e)

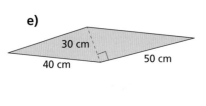

30 cm

40 cm

50 cm

2 Find the area of each triangle.

a)

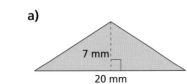

7 mm

20 mm

b)

15 km

4 km

c)

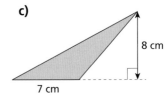

8 cm

7 cm

d)

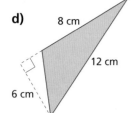

8 cm

12 cm

6 cm

e)

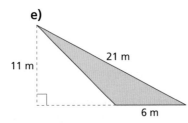

21 m

11 m

6 m

f)

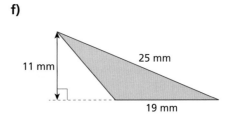

25 mm

11 mm

19 mm

3 Find the area of each trapezium.

a)

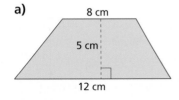

8 cm

5 cm

12 cm

b)

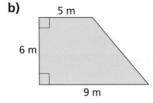

5 m

6 m

9 m

c)

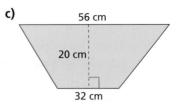

56 cm

20 cm

32 cm

4 Eleanor is calculating the area of this triangle:

She writes:

Area = $\frac{1}{2}$ (6 × 4) = 12 cm²

Do you agree with Eleanor? Explain your answer.

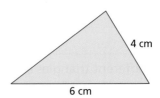

4 cm

6 cm

5 Which of these shapes has an area of 24 cm²? Explain how you know.

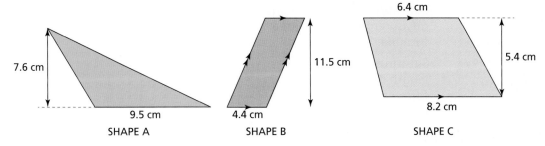

a)
2 cm
12 cm

b)
5 cm
4 cm
7 cm

c)
4 cm
12 cm

d)
3 cm
8 cm

6 Find which of these shapes has the largest area. Show how you decided.

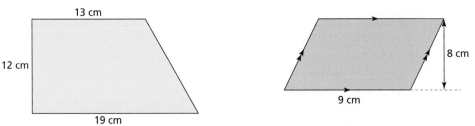

7.6 cm
9.5 cm
SHAPE A

11.5 cm
4.4 cm
SHAPE B

6.4 cm
5.4 cm
8.2 cm
SHAPE C

7 Show that the area of the parallelogram is $\frac{3}{8}$ of the area of the trapezium.

13 cm
12 cm
19 cm

8 cm
9 cm

11.3 Volume of triangular prisms

> **Think about**
>
> The volume of a cuboid can be found by the area of the cross-section multiplied by the length.
>
> Imagine cutting this cuboid in half to make the two triangular prisms, as shown.
>
> Can you see that the volume of each prism is the area of the triangular end multiplied by the length?
>
> The volume of a triangular prism = area of triangular cross-section × length

Worked example 3

Work out the volume of each solid.

a)
6 cm
8 cm
9 cm

b)
6 cm
10 cm
5 cm

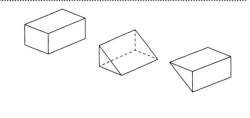

Volume = area of cross section × length		
a) Area of a triangle $= \frac{1}{2} \times$ base $\times$ height $= \frac{1}{2} \times 6 \times 8$ $= 24$ cm²	First work out the area of the cross-section of the prism. This is the area of the right angled triangle. Substitute base and height measurements in the area formula.	
Volume = 24 cm² × 9 $= 216$ cm³	Substitute the area and length into the volume formula: volume = area of cross-section × length.	
b) Area of a triangle $= \frac{1}{2} \times$ base $\times$ height $= \frac{1}{2} \times 10 \times 6$ $= 30$ cm² Volume = 30 cm² × 5 cm $= 150$ cm³	First work out the area of the cross-section of the prism. This is the area of the triangle. Substitute base and height measurements into the area formula. Substitute the area and length into the volume formula: volume = area of cross-section × length.	

Exercise 3

1 A cuboid has a volume of 72 cm³. All the dimensions of the cuboid are a whole number of centimetres. One side of the cuboid is 3 cm. List all the possible pairs of values of the other two sides. What can you say about each pair of values?

2 Find the volume of each of these triangular prisms.

a)

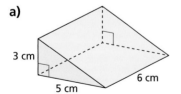

b)

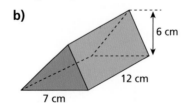

c)

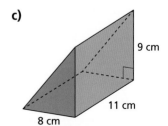

3 Find the difference in volumes of these two triangular prisms.

A

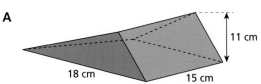

11 cm

18 cm   15 cm

B

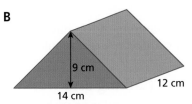

12 cm

14 cm   18 cm

4 Find which of these triangular prisms has the greatest volume.

A

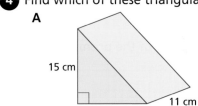

15 cm

13 cm   11 cm

B

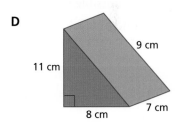

9 cm

14 cm   12 cm

C

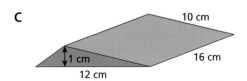

10 cm

1 cm   16 cm

12 cm

D

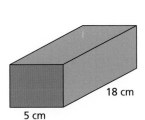

9 cm

11 cm

8 cm   7 cm

5 The triangular prism and the cuboid below have the same volume.
Calculate the height of the cuboid.

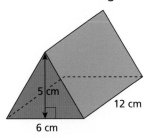

5 cm

12 cm

6 cm

18 cm

5 cm

6 Look at these triangular prisms. Oliver says one is 4 times the
volume of the other. Is Oliver correct? Give reasons for your answer.

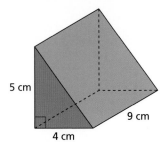

5 cm

9 cm

4 cm

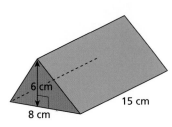

6 cm

8 cm

15 cm

7 The diagram shows a cube and a triangular prism. Show that the volume of the cuboid is 6 times larger than the volume of the prism.

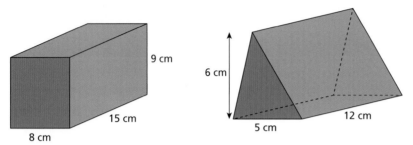

8 A triangular prism has a volume of 240 cm³.
It stands on a rectangular base measuring 8 cm by 6 cm. Find the height of the prism.

Thinking and working mathematically activity

A triangular prism has a volume of 90 cm³.

- Sketch one possible solution with integer side lengths.
- Can you find at least another five possible solutions?

11.4 Surface area of 3D shapes

Worked example 4

Here is the net of a square-based pyramid. Use the net to calculate the surface area of the pyramid.

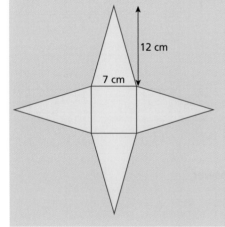

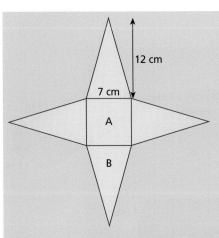

Area A = 7 × 7 = 49 cm²

Area B = $\frac{1}{2}$ (7 × 12)

$\quad$ = $\frac{1}{2}$ (84) = 42 cm²

Surface area of the pyramid

= Area A + 4(Area B)

= 49 + 4(42)

= 49 + 168

= 217 cm²

The surface area of the pyramid is the sum of the individual areas of each of the 5 faces.

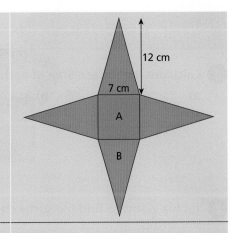

We know from the question text that the base of the pyramid is a square.

So area of base

= length × width

= 7 × 7

= 49 cm²

Because this is a pyramid, the 4 triangle faces will have equal lengths and hence equal areas.

The area of each triangle

= $\frac{1}{2}$ base × height

= $\frac{1}{2}$ (7 × 12)

= $\frac{1}{2}$ (84)

= 42 cm²

So the total surface area

= area of square base

$\quad$ + 4 × area of triangles

= 49 + 4 × 42

= 49 + 168

= 217 cm²

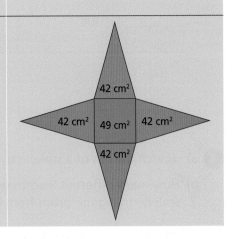

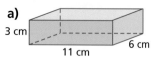

1 Calculate the surface area of each of these cuboids:

a)
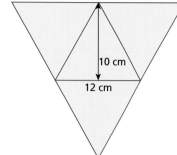
3 cm
11 cm
6 cm

b)
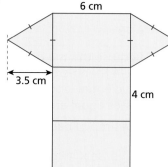
5 m
9 m
2 m

c)
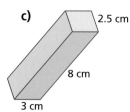
2.5 cm
8 cm
3 cm

d)
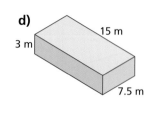
15 m
3 m
7.5 m

2 Sophia wants to find the surface area of this cuboid.
She writes:
surface area = 6 × area of one face = 6 × 60 = 360 cm²
Is Sophia correct? Give reasons for your answer.

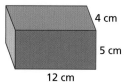
4 cm
5 cm
12 cm

3 Find the surface area of the shape made by each net.

a)

10 cm
12 cm

b)
6 cm
3.5 cm
4 cm

c)
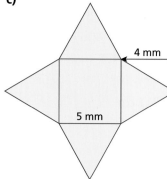
4 mm
5 mm

d)
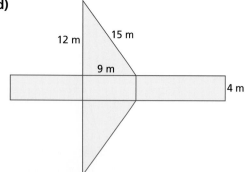
12 m
15 m
9 m
4 m

e)
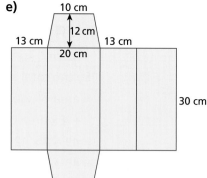
10 cm
12 cm
13 cm
13 cm
20 cm
30 cm

4 a) Sketch the net of a scalene triangular prism.

b) How many different lengths do you need to know to be able to calculate the surface area of a scalene triangular prism from its net? Explain your answer.

Thinking and working mathematically activity

A 3D shape has a surface area of 160 cm². All the dimensions of the shape are a whole number of centimetres.

- Sketch one possible solution with integer side lengths.
- Can you find another possible solution?

---

1 Which of the following distances is closest to 100 km?

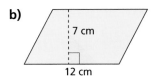

50 miles   60 miles   120 miles   160 miles

2 Eve drives 140 miles in the UK before crossing the channel on a ferry. She then drives 220 km in France. Work out the total distance she has driven in kilometres.

3 Find the area of each parallelogram.

a)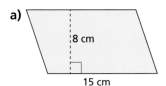

8 cm

15 cm

b)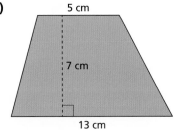

7 cm

12 cm

4 Find the area of each trapezium.

a)

7 cm

6 cm

3 cm

b)

5 cm

7 cm

13 cm

5 Find the unknown lengths in the following.

a) Find the length of *BE*.

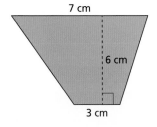

A

E

12 cm

6 cm

B

C

D

16 cm

b) Find the length of *DF*.

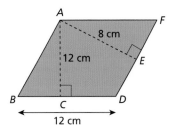

A

F

8 cm

12 cm

E

B

C

D

12 cm

6 Find the volume of each triangular prism.

a)

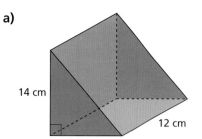

14 cm

12 cm

11 cm

b)

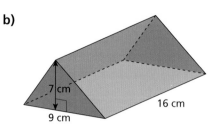

7 cm

9 cm

16 cm

7 Calculate the difference between the surface area of the cuboid and the prism formed by the net.

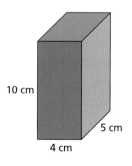

10 cm

5 cm

4 cm

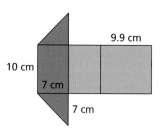

9.9 cm

10 cm

7 cm

7 cm

End of chapter reflection

You should know that...	You should be able to...	Such as...
A kilometre is approximately $\frac{5}{8}$ of a mile. A mile is approximately 1.6 kilometres	Convert between miles and kilometres.	a) Convert 5 miles to kilometres. b) Convert 24 kilometres to miles.
The area of a triangle $= \frac{1}{2}$ base × height $= \frac{1}{2} bh$ *h* *b*	Calculate the area of a triangle.	Find the area of the triangle. 6 cm 8 cm
The area of a parallelogram = base × perpendicular height = *bh* *h* *b*	Calculate the area of a parallelogram.	Find the area of this shape. 4 cm 11 cm

The area of a trapezium = average length × perpendicular height $= \frac{1}{2}(a + b)h$	Calculate the area of a trapezium.	Find the area of this shape.
The volume of a triangular prism = area of triangular cross-section × length	Calculate the volume of a triangular prism.	Find the volume of this solid.
The surface area of a 3D shape is the sum of the area of all its faces.	Calculate the surface area of shapes like triangular prisms and pyramids.	Find the total surface area of this solid.

12

Probability 1

You will learn how to:

- Understand that complementary events are two events that have a total probability of 1.
- Design and conduct chance experiments or simulations, using small and large numbers of trials. Compare the experimental probabilities with theoretical outcomes.

Starting point

Do you remember…

- the language of probability (likely, unlikely, certain, impossible)?

  For example, it is .... that I will be at school tomorrow.

- the probability scale from 0 to 1, where 0 means that an outcome is impossible and 1 means that an outcome is certain?

  For example, what is the probability that tomorrow is Sunday?

- how to find probabilities based on equally likely outcomes in simple contexts?

  For example, if a number is picked at random from the numbers 1, 2, 3, 4, 5 and 6, what is the probability that the number chosen is a 3?

- the notation P( ) means the probability of the outcome in the brackets?

  For example, P(3) means the probability of getting a 3.

- the word 'fair' means that a dice or spinner is equally likely to land on any side or face?

  For example, when a fair six-sided dice is rolled, the probability of it landing on each number is $\frac{1}{6}$

- how to identify all the possible mutually exclusive outcomes of a single event?

  For example, if a bag contains three red counters and four green counters and a counter is taken out of the bag at random, what are the possible outcomes?

- how to use experimental data to estimate probabilities?

  For example, a counter is taken out of a bag and then replaced. This is done a total of 100 times. A red counter is chosen 25 times. Calculate an estimate of the probability of choosing a red counter from the bag.

This will also be helpful when…

- you learn how to calculate the probability of an event not occurring, given the probability of the event occurring
- you find probabilities based on equally likely outcomes in practical contexts
- study probabilities for multiple events.

12.0 Getting started

Work in pairs with a set of 10 cards that are identical in size and colour, with a different number from 1 to 10 written on each card. You can either make the cards or your teacher will give them to you.

Place the cards face down and mix them up.

Now pick a card.

What is the probability that the card has the number 8 on it?

What is the probability that the card has a multiple of 5 on it?

Can you think of two questions which both have the answer $\frac{5}{10}$ or $\frac{1}{2}$?

Write down some probability questions of your own about picking a card from your 10 cards. Try to find questions with different answers.

12.1 Complementary events

Key terms

The **complement** of any event A is the event *not A*, written as A', and can be represented as $P(A) + P(A') = 1$

If the probability of an event occurring is p, then the probability of it *not* occurring is $1 - p$.

For example, if the probability of landing on a red section on a spinner is 0.3 then the probability of landing in any other colour (not landing on a red section) is 0.7

Worked example 1

a) The probability that a student chosen at random from class 8C owns a cat is $\frac{6}{10}$

What is the probability that a student chosen at random from class 8C does not own a cat?

b) The probability that a student chosen at random from class 8C owns a pair of grey trousers is 0.3

What is the probability that a student chosen at random from class 8C does not own a pair of grey trousers?

| a) P(does not own a cat) $= 1 - P(\text{owns a cat})$ $= 1 - \frac{6}{10}$ $= \frac{10}{10} - \frac{6}{10} = \frac{4}{10}$ | The probability that something does not happen is 1 – (the probability that it does happen). Change the 1 to $\frac{10}{10}$ to make the calculation easier. | |
| b) P(does not own grey trousers) $= 1 - P(\text{does own grey trousers})$ $= 1.0 - 0.3 = 0.7$ | Change the 1 to 1.0 to make the calculation easier. | |

Worked example 2

a) The probability that a student chosen at random from class 8C brings a packed lunch to school every day is 55%

What is the probability that a student chosen at random from class 8C does not bring a packed lunch to school every day?

b) All apartments in a block have either a green door or a red door. If the probability of an apartment chosen at random having a red door is $\frac{3}{7}$, what is the probability that an apartment chosen at random has a green door?

| **a)** P(does not bring packed lunch) = 1 − P(does bring packed lunch)

 = 100% − 55% = 45% | When you are dealing with percentages, remember that

 $1 = \frac{100}{100} = 100\%$ | |
| **b)** P(a green door)

 = P(not a red door)

 = 1 − P(red door)

 = $1 - \frac{3}{7} = \frac{4}{7}$ | As the doors can only be red or green then the probability of a green door is the same as the probability of not having a red door. | |

Exercise 1

1 The probability that the school bus arrives on time on any Friday is $\frac{1}{3}$. What is the probability that the school bus will not arrive on time next Friday?

2 The probability of a student chosen at random from class 8A owning a pet is $\frac{5}{9}$. What is the probability of a student chosen at random from class 8A not owning a pet?

3 Jana picks a sweet at random from a box of sweets, of identical shape and size, that are all either orange or lemon sweets. If the probability that she picks an orange sweet is 0.2, what is the probability that she picks a lemon sweet?

4 In Marco's sock drawer there are only black socks and red socks. If Marco picks a sock at random, the probability that the sock will be black is 0.63. What is the probability that Marco picks out a red sock?

5 The probability that a member of Marika's class gets the bus to school is 53%. What is the probability that a member of Marika's class, chosen at random, does not get the bus to school?

6 This table shows the probability that it will rain on each day of a given week.
Copy and complete the table.

Event	Monday	Tuesday	Wednesday	Thursday	Friday
P(rain)	0.5	0.88	0.95	0.72	0.15
P(no rain)					

7 There are 100 batteries, of identical shape and size, in a box.

The probability that a battery chosen at random is not flat is 0.77

How many flat batteries are in the box?

8 There are 80 sweets, of identical shape and size, in a jar. The sweets are either mints or toffees. The probability that a sweet chosen at random is a mint is 40%. How many toffees are in the jar?

9 A bag holds a number of counters, of identical shape and size, which are either red, yellow or blue. The probability that a counter chosen at random is red is 0.2. The probability that a counter chosen at random is yellow is 0.35.

What is the probability that a counter chosen at random is blue?

10 The probability that the school bus will be late on a morning when it is raining is 0.75.

Hanna says that this means the probability the bus will be late on a morning when it is not raining is 0.25.

Explain, with reasons, why Hanna is not correct and write a correct statement from the information given.

> **Discuss**
>
> What is the smallest number of counters that could be in the bag in question 9? Discuss as a class.

 Thinking and working mathematically activity

A bag contains balls that are either red or yellow or green. When a ball is picked at random:

- the probability that it is not red is $\frac{7}{12}$

- the probability that it is not yellow is $\frac{5}{8}$

Decide if the following statements must be true, could be true or must be false. Give a reason for each answer.

- The number of green balls is 12
- The smallest possible number of red balls in the bag is 10
- The total number of balls in the bag is 48
- There are twice as many red balls as yellow balls.

Write a statement that must be true about the total number of balls in the bag.

12.2 Experimental probability

Key terms

Experimental probability is an estimate of the probability of a particular outcome of an event based on the outcomes of several repetitions of the event. This is also known as **relative frequency**. The formula is:

Experimental probability = relative frequency = $\dfrac{\text{number of successful trials}}{\text{total number of trials}}$

Theoretical probability is defined by the formula:

Theoretical probability = $\dfrac{\text{number of favourable events}}{\text{total number of possible outcomes}}$

Worked example 3

Anna is going to roll a six-sided dice. She does not know whether or not it is fair.
After rolling the dice 30 times, the results are:

Dice roll	1	2	3	4	5	6
Frequency	2	3	8	3	10	4

a) What is the theoretical probability of rolling each number on a six-sided dice?

b) What is the experimental probability for each number, based on Anna's rolls?

c) Do you think the dice is fair?

a) $P(1) = P(2) = P(3) = P(4) = P(5) = P(6) = \frac{1}{6}$

The theoretical probability of rolling each number on a six-sided dice is the number of ways each number can occur divided by the number of possible outcomes.

b)

Roll	Experimental probability
1	$\frac{2}{30} = \frac{1}{15}$
2	$\frac{3}{30} = \frac{1}{10}$
3	$\frac{8}{20} = \frac{4}{15}$
4	$\frac{3}{30} = \frac{1}{10}$
5	$\frac{10}{30} = \frac{1}{3}$
6	$\frac{4}{30} = \frac{2}{15}$

To find the experimental probability you work out:

$$\frac{\text{number of successful trials}}{\text{total number of trials}}$$

c) The experimental probabilities are quite different from the theoretical probabilities, so this suggests that the dice is not fair.

You can see if the dice is fair by comparing the theoretical and experimental probabilities.

Discuss

How could Anna be more confident with her decision about whether or not the dice is fair?

Did you know?

If you flip a coin 100 times, the probability that you will get fifty heads and fifty tails is less than 0.08

Exercise 2

1 Charlie's results from rolling a four-sided dice are shown in the table.

a) What is the theoretical probability that each number will be rolled?

b) What is the experimental probability that each number will be rolled?

c) Is Charlie's dice fair? Explain your answer.

Charlie				
Number	1	2	3	4
Frequency	10	10	9	11

2 Henry and Mason each roll a four-sided dice. Their results are shown in the tables.

Henry				
Number	1	2	3	4
Frequency	11	14	10	15

Mason				
Number	1	2	3	4
Frequency	33	17	28	22

a) How many times did they each roll their dice?

b) What is the theoretical probability that each number will be rolled for each dice?

c) Find the relative frequency of rolling each number on both Henry's and Mason's dice.

d) Whose dice has results closer to those expected for theoretical probability? Whose dice is fairer? Explain your answer.

3 Akong rolls a six-sided dice. His results are shown in the table.

a) What is the theoretical probability that each number will be rolled?

b) Find the relative frequency of rolling each number on Akong's dice.

c) Is Akong's dice fair? Explain your answer.

Akong						
Number	1	2	3	4	5	6
Frequency	11	12	9	10	9	10

4 Abdo's results from rolling a six-sided dice are shown in the table.

a) What is the theoretical probability that each number will be rolled?

b) What is the experimental probability that each number will be rolled?

c) Is Abdo's dice fair? Explain your answer.

Abdo						
Number	1	2	3	4	5	6
Frequency	10	7	7	11	10	15

5 a) What is the theoretical probability of a fair dice showing a six when it is thrown?

b) A dice is thrown 75 times and shows a six 25 times. What is the experimental probability of throwing a six with this dice?

c) Do you think that the dice in part **b)** is a fair dice? Explain your answer.

6 Talia and Favor's results from rolling a six-sided dice are shown in the tables.

Talia						
Number	1	2	3	4	5	6
Frequency	15	11	14	16	13	11

Favor						
Number	1	2	3	4	5	6
Frequency	20	19	11	15	11	4

a) What is the theoretical probability that each number will be rolled?

b) What is the experimental probability that each number will be rolled?

c) Whose dice is fairer? Explain your answer.

7 Tariq spins a spinner with four equal sectors. His results are shown in the table.

Colour	Red	Blue	Green	Yellow
Frequency	17	20	13	30

Tariq says, 'The spinner is fair because the experimental probability for blue is equal to $\frac{1}{4}$'.
Is Tariq correct? Explain your answer.

8 A spinner has four different coloured sections. The spinner is spun 200 times and the colour on which it lands is noted.

a) Copy and complete the table.

	Red	Purple	Blue	Black
Frequency	36		72	
Experimental probability		0.29	0.36	

b) Which two colour sections are likely to be the same size? Explain your reasoning.

> **Discuss**
>
> A fair spinner has 6 sections, each a different colour. Jana spins the spinner 100 times and it lands on the red section 18 times. Sara spins the same spinner 100 times. Will she also find that the spinner lands on red exactly 18 times?

> **Think about**
>
> How many times would you need to roll a dice to decide if it was fair or biased?

 Thinking and working mathematically activity

Create a four-sided spinner.

Design an experiment to test if your spinner is fair.

Present your results in a suitable way and make a conclusion.

Consolidation exercise

1 State whether or not the following are complementary events:

a) A team wins a game of football
A team loses a game of football

b) It will rain tomorrow
It will not rain tomorrow.

c) I will pass a test
I will fail the test.

d) Getting an even number when rolling a dice
Getting a prime number when rolling a dice

2 The probability that Javid cycles to school is $\frac{2}{5}$ and the probability that he walks to school is $\frac{1}{5}$. What is the probability that Javid doesn't walk or cycle to school?

3 There are 100 counters in a bag. The probability of picking a red counter is 0.45

a) What is the probability of not picking a red one?

b) How many counters are not red?

4 Jake and Peter each spin a spinner 20 times. The results are shown in the table below.

	Lands on red	Lands on blue	Lands on green
Jake	6 times	6 times	8 times
Peter	7 times	8 times	5 times
Combined total			

For each of the following statements, say whether it is true, maybe true or false. If it is false or maybe true explain your reasoning.

a) The estimated experimental probability of Jake's spinner landing on blue is 0.3

b) The estimated experimental probability of Peter's spinner landing on blue is 0.4

c) If you combine the two sets of results then the estimated experimental probability of the spinner landing on blue is 0.7

d) The spinner is equally likely to land on any of the three colours.

e) There are only three colours on the spinner.

5 Sara rolls a six-sided dice 60 times.

Copy and complete the table below:

Number shown on dice	1	2	3	4	5	6
Number of times	8	10	9	13	9	11
Theoretical probability	$\frac{1}{6}=\frac{\square}{60}$	$\frac{1}{6}=\frac{\square}{60}$				
Relative frequency			$\frac{9}{60}$			

End of chapter reflection

You should know that...	You should be able to...	Such as...
The complement of any event A is the event *not A*, written as A', If the probability that an event occurs is p, then the probability that it does not occur is $1 - p$.	Calculate the probability that an event does not occur when you are given the probability of the event occurring.	If there is a 23% probability of rain tomorrow, what is the probability that it will not rain?
Experimental probability/relative frequency uses real experimental data $P(\text{event}) = \dfrac{\text{Number of times the event occurred}}{\text{Number of trials}}$	Use experimental data to estimate probabilities.	List the experimental probabilities of landing on each colour for this spinner from this table of results: <table><tr><td>Colour</td><td>red</td><td>blue</td><td>green</td></tr><tr><td>spins</td><td>12</td><td>16</td><td>12</td></tr></table>
Theoretical probability requires every outcome to be mutually exclusive and equally likely. $P(\text{event}) = \dfrac{\text{Number of ways the event could occur}}{\text{Number of possible outcomes}}$	Compare experimental and theoretical probabilities in simple contexts.	These are Victoria's results when she rolled a four-sided dice. Is her dice a fair dice? Give a reason for your answer. <table><tr><td>Number</td><td>1</td><td>2</td><td>3</td><td>4</td></tr><tr><td>Frequency</td><td>9</td><td>19</td><td>14</td><td>8</td></tr></table>

13

Calculations

You will learn how to:

- Understand that brackets, indices (square and cube roots) and operations follow a particular order.
- Use knowledge of the laws of arithmetic and order of operations (including brackets) to simplify calculations containing decimals or fractions.

Starting point

Do you remember...

- how to apply order of operations to calculations that include brackets, positive indices and the four operations?

  For example, find $(2 + 3^2) \times 5$

- how to use laws of arithmetic and order of operations to simplify calculations (without brackets) containing decimals or fractions?

  For example, simplify and calculate: $2.2 + 3.9 - 0.2$, $2.5 \times 17 \times 4$

This will also be helpful when...

- you use order of operations with negative indices and with more complicated calculations
- you use order of operations, inverse operations and equivalence to simplify calculations containing both decimals and fractions
- you use order of operations to simplify calculations containing fractions and decimals.

13.0 Getting started

Find $(3 + 2)^2$ and $3^2 + 2^2$. Are the answers the same or different? Explain this using order of operations.

Explain how the diagrams below represent $(3 + 2)^2$ and $3^2 + 2^2$.

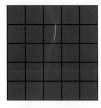 

Find $(4 - 2)^2$ and $4^2 - 2^2$. Are the answers the same or different? Explain this using order of operations.

Explain how the diagrams below represent $(4 - 2)^2$ and $4^2 - 2^2$.

Use a calculator to do the calculations below. Make sure that in $\sqrt{9+16}$, the whole addition is underneath the square root sign, or else use brackets like this: $\sqrt{(9+16)}$. Do the same for $\sqrt{25-9}$. In each pair, are the answers the same or different? Suggest why.

$$\sqrt{9+16} \text{ and } \sqrt{9} + \sqrt{16}$$
$$\sqrt{25-9} \text{ and } \sqrt{25} - \sqrt{9}$$

Draw a diagram to represent each calculation.

13.1 Order of operations

Key terms

The **order of operations** is the agreed order of calculating in mathematics. The order is:

1. Brackets
2. Indices (powers and roots)
3. Division and Multiplication (from left to right)
4. Addition and Subtraction (from left to right)

The $\sqrt{}$ symbol tells you to find the positive square root of a number. So $\sqrt{9}$ means 3, not −3.

> **Did you know?**
>
> These expressions all have the same meaning:
>
> $$\sqrt{(9+16)} \qquad \sqrt{(9+16)} \qquad \sqrt{9+16}$$
>
> They tell you to do the addition first and then find the square root of the result. The answer is 5.
> These expressions have the same meaning but it is a different meaning to those above:
>
> $$\sqrt{9} + 16 \qquad \sqrt{9} + 16$$
>
> The square root only applies to the 9, so find $\sqrt{9}$ first and then add 16 to the result. The answer is 19.

Worked example 1

Find:

a) $10 - (12 + (5 - 3))$ b) $\sqrt{8^2 + 36} \div 2$ c) $\dfrac{20 - \sqrt[3]{8}}{2 \times 3}$

a) $10 - (12 + (5 - 3)) = 10 - (12 + 2)$ $= 10 - 14$ $= -4$	When there are brackets inside brackets, start with the calculation in the innermost brackets. Do the calculation in the outer brackets. Then do the subtraction.

$\sqrt{8^2 + 36} \div 2 = \sqrt{64 + 36} \div 2$ $= \sqrt{100} \div 2$ $= 10 \div 2$ $= 5$	Treat calculations under a square root sign as if they are in brackets: $\sqrt{8^2 + 36} = \sqrt{8^2 + 36}$ First, do the squaring. Then do the addition. Do the square rooting. Do the division.
$\dfrac{20 - \sqrt[3]{8}}{2 \times 3} = \left(20 - \sqrt[3]{8}\right) \div (2 \times 3)$ $= (20 - 2) \div 6$ $= 18 \div 6$ $= 3$	Treat expressions in a fraction as if they are in brackets. Do the calculations in brackets first. In the first set of brackets, do the cube rooting before the subtracting. Then do the division.

Exercise 1

1-2, 4–8

1 Find:

a) $10 - 5 \times 4$

b) $(10 - 5) \times 4$

c) $10 \div 2 + 3$

d) $10 \div (2 + 3)$

e) $3 + 5 \times \sqrt{16}$

f) $\sqrt{4} \times (2 - 5)$

g) $6 \times \sqrt[3]{8} \div 2$

h) $\sqrt{2 \times 8} \div 4$

2 Find:

a) $36 \div 3^2 - 2^2$

b) $12 - (12 - 7)^2$

c) $\sqrt{3^2 \times 4^2}$

d) $\sqrt{10^2 - 8^2} \times 5$

e) $4 \times (9 - (3 + 4))$

f) $10 + \sqrt[3]{20 - 12}$

g) $4 \times 5^2 \div (3 + 2)^2$

h) $\sqrt[3]{6 - 2 \times 7}$

i) $\sqrt{43 + (3 \times 2)}$

3 **Technology question** Use a calculator to find:

a) $(7 - \sqrt[3]{64}) \times 3$

b) $\dfrac{18 - 6 \div 3}{\sqrt{16}}$

c) $(3 - 5)^2 \times \sqrt[3]{27}$

d) $\sqrt[3]{12 - 4} \div 4$

e) $\dfrac{3 + 20 \div 2^2 \times 6}{\sqrt{9}}$

f) $\dfrac{\sqrt{7 \times 4 + 8}}{2}$

g) $2^2 \times \sqrt[3]{5^2 + 10^2}$

h) $\sqrt[3]{3^3 + 4^3 + 5^3}$

4 Dominic wants to find $\sqrt{9 + 16}$

He says, 'In the correct order of operations, square roots come before addition.'
His working is below.

$\sqrt{9 + 16} = \sqrt{9} + \sqrt{16}$

$= 3 + 4$

$= 7$

State whether or not his calculation is correct. Explain your answer.

5 Use the numbers 9, 11, 20 and 45 once each to make this calculation correct.

($\blacklozenge$ + $\blacklozenge$) ÷ ($\blacklozenge$ − $\blacklozenge$) = 6

6 Find the missing numbers.

a) 5^2 + ($\blacksquare$ − 3 × 5) = 100

b) (29 − $\sqrt{\blacksquare}$ × 2) ÷ 3 = 7

7 Insert brackets to make the statements correct.

a) 48 + 12 ÷ 4 × 1 + 2 = 50

b) 48 + 12 ÷ 4 × 1 + 2 = 49

c) 48 + 12 ÷ 4 × 1 + 2 = 57

8 Insert square root signs to make the statements correct.

a) 9 + 16 × 4 ÷ 2 = 35

b) 9 + 16 × 4 ÷ 2 = 25

c) 9 + 16 × 4 ÷ 2 = 10

d) 9 + 16 × 4 ÷ 2 = 17

e) 9 + 16 × 4 ÷ 2 = 13

f) 9 + 16 × 4 ÷ 2 = 7

Think about

You can write one square root inside another. Calculate the innermost root first. For example,
$\sqrt{22 + \sqrt{7 + 2}} = \sqrt{22 + 3} = \sqrt{25} = 5$.

By writing two square root signs in 9 + 16 × 4 ÷ 2, with one root inside the other, make the answer 5. Find a different answer using one square root sign inside another.

Discuss

Brackets and the order of operations are very important in real life. Suppose two classes of 18 and 13 children are going to visit the Blue Mountains in Australia on a school trip. The teacher takes two bottles of water for each child. The calculation is 2 × (18 + 13) = 62 bottles. What mistakes could the teacher make if there were no brackets? Think about other examples where brackets and order of operations are important in real life.

 Thinking and working mathematically activity

Use +, −, × and ÷ to make the calculations correct.

$\sqrt{6 \, \blacksquare \, 6}$ ÷ 3 = 2

(2^3 $\blacksquare$ 28) ÷ 3^2 = 4

9 $\blacksquare$ $\sqrt[3]{25 + 2}$ = 3

(7 × 2 − 2 $\blacksquare$ 3)^2 = 64

(50 $\blacksquare$ 6^2) ÷ (2 $\blacksquare$ 5) = 2

3 × $\sqrt{18 \, \blacksquare \, 2}$ × 2^2 + 8^2 = 10^2

Discuss your strategies for solving these problems.
Write your own questions like these.

Key terms

The **associative** law says that when you add or multiply numbers, you can group them in different ways without changing the result.

$(a + b) + c = a + (b + c)$

$\left(\frac{2}{9} + \frac{1}{8}\right) + \frac{7}{8} = \frac{2}{9} + \left(\frac{1}{8} + \frac{7}{8}\right)$

$(a \times b) \times c = a \times (b \times c)$

$(3.1 \times 4) \times 0.5 = 3.1 \times (4 \times 0.5)$

Worked example 2

Do these calculations as efficiently as you can.

a) $0.25 \times 1.52 \times 4$

b) $6\frac{3}{4} + \left(\frac{1}{4} + \frac{7}{9}\right)$

c) 9.9×1.8

d) $0.4 \times 28 + 0.6 \times 28$

a) $0.25 \times 1.52 \times 4 = 0.25 \times 4 \times 1.52$ $= 1 \times 1.52$ $= 1.52$	Swap numbers so the calculation starts with 0.25×4. This gives you 1, which is easy to work with.	With multiplication $a \times b = b \times a$. This is the commutative law.
b) $6\frac{3}{4} + \left(\frac{1}{4} + \frac{7}{9}\right) = \left(6\frac{3}{4} + \frac{1}{4}\right) + \frac{7}{9}$ $= 7 + \frac{7}{9}$ $= 7\frac{7}{9}$	Group the numbers in a different way, so that the first calculation is $6\frac{3}{4} + \frac{1}{4}$. This gives you 7, which is easy to work with.	Use the associative law to make the addition easier without changing the answer. 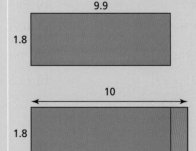
c) $9.9 \times 1.8 = (10 - 0.1) \times 1.8$ $= 10 \times 1.8 - 0.1 \times 1.8$ $= 18 - 0.18$ $= 17.82$	Partition 9.9 into $10 - 0.1$ Then use the distributive law to make the multiplication easier without changing the answer.	

d) $0.4 \times 28 + 0.6 \times 28 = (0.4 + 0.6) \times 28$
$$= 1 \times 28$$
$$= 28$$

The number 28 appears in both multiplications.

Factorise by taking 28 outside the brackets.

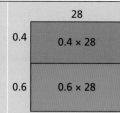

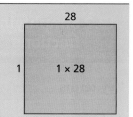

Exercise 2

1 Which pairs of calculations give the same answer?

A $(161 \times 5) \times 2$ $161 \times (5 \times 2)$

B $48 \times 1\frac{3}{4}$ $48 \times 2 - 48 \times \frac{1}{4}$

C $(17 \times 29) - (7 \times 29)$ 24×29

D $0.25 \div 4$ $4 \div 0.25$

E $(3.1 \times 5.4) \times 0.9$ $3.1 \times (5.4 \times 0.9)$

F $0.25 + 0.37 + 0.75$ $0.75 + 0.25 + 0.37$

G $4 \div \left(\frac{4}{9} \div \frac{2}{3}\right)$ $\left(4 \div \frac{4}{9}\right) \div \frac{2}{3}$

H $\frac{7}{8} + \left(\frac{1}{8} + \frac{2}{3}\right)$ $\left(\frac{7}{8} + \frac{1}{8}\right) + \frac{2}{3}$

> **Tip**
>
> It is possible to answer this question without doing the calculations.

2 Use efficient methods to find the answers. Show your working.

a) $87 + (13 + 128)$
b) $1.5 + (2.5 + 3.9)$
c) $175 + 61 + 25$
d) $7.6 + 3.89 + 2.4$
e) $(3.71 + 12.25) + 0.75$
f) $4.2 + 6.7 + 5.8 + 7.3$

3 Use efficient methods to find the answers to the calculations below. Show your working.

If an answer is a fraction greater than 1, write it as a mixed number.

a) $\left(\frac{5}{8} + \frac{4}{7}\right) + \frac{3}{7}$
b) $\frac{2}{3} + \frac{5}{12} + \frac{1}{3}$
c) $\left(\frac{1}{2} + \frac{7}{12}\right) + \frac{11}{12}$
d) $1\frac{1}{2} + \frac{7}{8} + \frac{1}{2} + \frac{1}{8}$
e) $\frac{2}{5} + \left(\frac{3}{5} + 1\frac{7}{11}\right)$
f) $2\frac{3}{4} + \left(1\frac{1}{4} + \frac{3}{5}\right) + \frac{1}{5}$

4 Find efficient ways of doing these calculations. Show your working.

a) 101×23
b) 99×0.6
c) 1.1×18
d) 0.9×6.7
e) 0.24×1.1
f) 0.6×9.9

5 Simplify the calculations where possible, and find the answers. Show your working.

a) $4 \times (25 \times 72)$
b) $(1.5 \times 0.7) \times 6$
c) $0.25 \times (1.8 \times 4)$
d) $0.5 \times (0.2 \times 15)$
e) $1.2 \times (0.7 - 0.3)$
f) $0.8 \times 3.2 \div 8$

6 Find:

a) $24 \times \dfrac{5}{8} \times 8$

b) $\dfrac{2}{3} \times 11 \times 9$

c) $\dfrac{2}{3} \times \left(1\dfrac{1}{2} - \dfrac{3}{4}\right)$

d) $\dfrac{5}{8} \div \left(\dfrac{1}{4} + \dfrac{1}{8}\right)$

e) $\dfrac{3}{5} \times \dfrac{1}{6} \div 2$

f) $\left(\dfrac{3}{4} \times \dfrac{5}{7}\right) \times 7$

g) $\dfrac{7}{10} - \dfrac{3}{5} \times \dfrac{1}{3}$

h) $2\dfrac{1}{2} \times 7 \times 2$

i) $\left(\dfrac{1}{12} + \dfrac{1}{4}\right) \times \dfrac{3}{5}$

7 Jane answered the question below correctly, but she used a long method.

Find the answer using an easier method.

$$49 \times \frac{7}{10} - 19 \times \frac{7}{10} = \frac{49 \times 7}{10} - \frac{19 \times 7}{10}$$
$$= \frac{343}{10} - \frac{133}{10}$$
$$= \frac{210}{10}$$
$$= 21$$

> **Think about**
>
> Write two methods for finding $5\dfrac{1}{3} \times \dfrac{5}{6}$ and two methods for finding 4.1×2.1. For each calculation, decide whether you find one of the methods easier than the other.

Thinking and working mathematically activity

Investigate which types of calculation obey the associative law. Explore calculations with three numbers and two operations.

Test different combinations of operations, which could be either the same, for example:

Does $(a \div b) \div c$ equal $a \div (b \div c)$?

or different, for example:

Does $(a \times b) + c$ equal $a \times (b + c)$?

Summarise your findings. Try to write a general rule and explain it.

Consolidation exercise

1 Amy works out $24 \div (2 + 4) = 18$. She does the following calculations:

$24 \div 2 = 12$

$24 \div 4 = 6$

$12 + 6 = 18$

Explain why Amy is wrong.

2 Gabrielle's garden is a square shape with sides of length 9.8 m. The garden is to be sown with grass seeds. The grass seeds cost \$8 per m² plus \$10 for delivery. The gardener estimates that he will need $(10 + 9.8^2) \times 8 \approx \$880$. Gabrielle estimates the cost to be $9.8 \times 8 + 10 \approx \$90$.

Do you agree with the estimates? Explain your answer.

3 Explain the difference between $\sqrt{25 - 9}$ and $\sqrt{25} - 9$

4 Find:

a) $2 \times (\sqrt{16} - 1)^2$

b) $\dfrac{27}{9 - 2^3}$

c) $2 - 1 \times 3^2$

d) $\sqrt[3]{3 \times 9} \times 8$

e) $100 - (10 - (3 + 1))^2$

f) $\sqrt{64} \div \sqrt[3]{8}$

g) $50 \times 4 - 7 \times 10^2$

h) $\sqrt[3]{4 \times 9 - (17 - 8)}$

5 Insert a cube root sign to make each calculation correct.

a) $125 - 27 + 8 = -14$

b) $128 - 8^2 - 27 = 97$

6 Insert one set of brackets to make each calculation correct.

a) $2^2 \times 5 + 5 \times 6 = 140$

b) $2^2 \times 5 + 5 \times 6 = 240$

c) $2^2 \times 5 + 5 \times 6 = 150$

7 Decide whether each statement is correct or incorrect. You can decide without actually performing the calculations.

a) $(4.5 \times 6) \times 8.2 = 4.5 \times (6 \times 8.2)$

b) $5 \times 35 - \dfrac{1}{5} \times 35 = 4\dfrac{4}{5} \times 35$

c) $(7.2 \div 2) \div 0.4 = 7.2 \div (2 \div 0.4)$

d) $(9.5 + 3.7) - 2.8 = 9.5 + (3.7 - 2.8)$

e) $\dfrac{1}{6} + \left(\dfrac{2}{3} + \dfrac{7}{9}\right) = \left(\dfrac{1}{6} + \dfrac{2}{3}\right) + \dfrac{7}{9}$

f) $\dfrac{3}{4} \times \left(\dfrac{1}{5} + \dfrac{3}{10}\right) = \dfrac{3}{4} \times \dfrac{1}{5} + \dfrac{1}{5} \times \dfrac{3}{10}$

8 A class of children make paper chain decorations. They make the following lengths:

- 15 chains are $\dfrac{2}{5}$ of a metre long
- 15 chains are $\dfrac{1}{2}$ a metre long

Write two methods for finding the total length of the paper chains.
Find the total length and write it as a mixed number.

9 A cuboid is 5 m long, 0.24 m wide and 0.8 m high. Find its volume.

10 Greg buys four items for $14.99 each. Use an efficient method to find the total cost.

End of chapter reflection

You should know that...	You should be able to...	Such as...
In the order of operations, square and cube roots have the same priority as indices. Treat calculations under a square or cube root sign like calculations in brackets: do them before other calculations.	Use the correct order of operations with calculations including square and cube roots.	Find: a) $2 \times (6 - \sqrt{2^2 + 5})$ b) $\sqrt[3]{8^2} \div 4$
Some calculations can be simplified by finding an efficient order to do the calculations.	Use the laws of arithmetic and order of operations to simplify calculations with fractions and decimals, including calculations with brackets.	Simplify and calculate: a) $6.88 + (1.12 + 3.97)$ b) $\left(\dfrac{7}{8} + 1\dfrac{2}{5}\right) + \dfrac{3}{5}$ c) $1\dfrac{3}{7} \times 8 + 8 \times \dfrac{4}{7}$ d) $3.65 \times 12 - 1.65 \times 12$

14

Equations and inequalities

You will learn how to:

- Understand that a situation can be represented either in words or as an equation. Move between the two representations and solve the equation (integer or fractional coefficients, unknown on either or both sides).
- Understand that letters can represent open and closed intervals (two terms).

Starting point

Do you remember...

- how to solve an equation with an unknown on one side?

  For example, solve $3x + 5 = 32$ or $22 = 30 - 2x$

- how to expand brackets?

  For example, expand $3(2x - 5)$

- how to use inequality symbols?

  For example, write "x is greater than or equal to 7" using mathematical symbols.

This will also be helpful when...

- you learn how to solve pairs of equations with two unknowns (called simultaneous equations)
- you learn to calculate bounds of accuracy.

14.0 Getting started

Look at rectangle A, below left.

- Can you find a value of x for which $2(x + 4) = 3x - 7$?
- What must the length of the rectangle be if this is the value of x?
- Are there are any other values of x that make the lengths equal?
- Now answer the previous three questions for rectangle B.
- Can you make up an example where there is no solution?

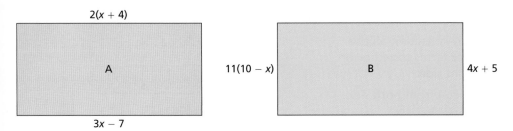

14.1 Solving equations with brackets and fractions

Worked example 1

Solve $3(3x + 2) = 42$

$3(3x + 2) = 42$	Start by expanding the brackets to create an equivalent equation. Multiply everything in the brackets by 3.	
$3 \times 3x + 3 \times 2 = 42$ $9x + 6 = 42$	Collect terms to simplify the equation.	
$9x + 6 = 42$ $\underline{-6 \quad -6}$ $9x = 36$	Subtract 6 from both sides.	
$9x = 36$ $\underline{\div 9 \quad \div 9}$ $x = 4$	Finally, divide both sides by 9 to find the value of x.	

Worked example 2

Solve $\dfrac{4}{5}y + 4 = 14$

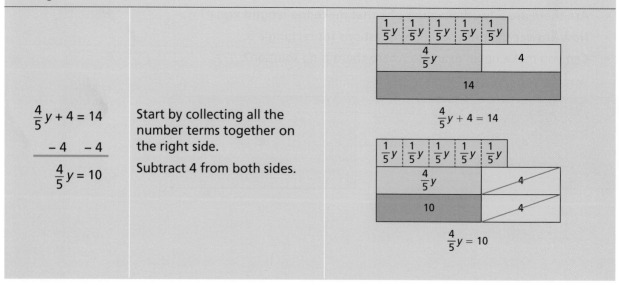

$\dfrac{4}{5}y + 4 = 14$ $\underline{-4 \quad -4}$ $\dfrac{4}{5}y = 10$	Start by collecting all the number terms together on the right side. Subtract 4 from both sides.	

$\frac{4}{5}y = 10$ $\div 4 \quad \div 4$ $\frac{1}{5}y = 2.5$	Divide by 4.	$\begin{array}{\|c\|c\|c\|c\|c\|}\hline \frac{1}{5}y & \frac{1}{5}y & \frac{1}{5}y & \frac{1}{5}y & \frac{1}{5}y \\\hline \frac{1}{5}y & \frac{1}{5}y & \frac{1}{5}y & \frac{1}{5}y \\\hline 2.5 & 2.5 & 2.5 & 2.5 \\\hline \end{array}$ $\frac{1}{5}y = 2.5$
$\frac{1}{5}y = 2.5$ $\times 5 \quad \times 5$ $y = 12.5$	Multiply by 5.	$\begin{array}{\|c\|}\hline y \\\hline 12.5 \\\hline \end{array}$ $y = 12.5$

Exercise 1

1 Solve these equations.

a) $4(x + 2) = 32$ b) $5(x - 3) = 25$ c) $2(3x + 1) = 14$ d) $3(5x + 10) = 90$

e) $7(2x - 1) = 77$ f) $3(3x - 2) = 75$ g) $24 = 2(x + 3)$ h) $54 = 9(5x - 19)$

2 Sofia and Thomas are solving the equation $3(5x - 1) = 42$

Here are their workings out:

Sofia

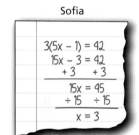

Thomas

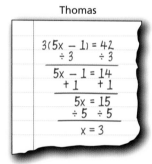

a) What has Thomas done differently to Sofia? Is his method correct?

b) Does this method always work?

c) Which method is better to use for an equation like $9(6x - 5) = 63$?

3 Solve these equations.

a) $\frac{2}{3}x = 8$ b) $\frac{3}{7}t = 15$ c) $\frac{2}{5}r = 0.6$ d) $\frac{7}{4}x = 42$

e) $\frac{1}{4}y + 6 = 9$ f) $\frac{2}{9}g - 5 = 11$ g) $\frac{3}{2}x + 15 = 3$ h) $\frac{4}{3}t - \frac{5}{12} = \frac{1}{4}$

4 Solve $3(x + 2) + 5(x + 4) = 90$

5 Solve these equations.

a) $8x - 3(x - 2) = 41$

b) $5(2y - 2) - 4(y - 1) = 48$

> **Tip**
>
> $-3(x - 2)$ expands to give $-3x + 6$

Thinking and working mathematically activity

Write down four equations that have $x = 3$ as a solution.

Your equations must include brackets, and an addition sign (+) or a subtraction sign (−).

14.2 Solving equations with unknowns on both sides

Worked example 3

Solve $1 + 2x = 5x - 8$

$1 + 2x = 5x - 8$ $\underline{-2x \quad\quad -2x}$ $1 = 3x - 8$	We must do the same to each side of the equation. First, subtract $2x$ from each side (or each row in the bar model).	The bar model has $1 + 2x$ on the row the same length as $5x - 8$ underneath: 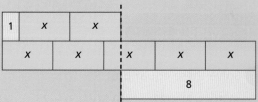
$1 = 3x - 8$ $\underline{+8 \quad\quad +8}$ $9 = 3x$	Add 8 to both sides (rows in the bar model). We now have an equation with an unknown on just one side.	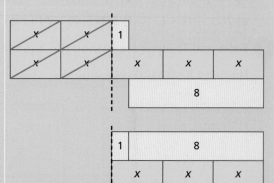
$9 = 3x$ $\underline{\div 3 \quad \div 3}$ $3 = x$	Finally, divide both sides (rows) by 3.	so $x = 3$

Worked example 4

Solve $2x + 3 = 7 - 3x$

$2x + 3 = 7 - 3x$ $\underline{+\ 3x \qquad\quad +\ 3x}$ $5x + 3 = 7$	Collect all unknowns onto the same side of the equation. Add $3x$ to both sides.	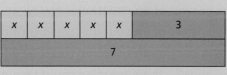 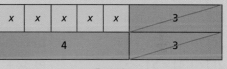 $5x + 3 = 7$
$5x + 3 = 7$ $\underline{-\ 3 \quad\ -\ 3}$ $5x = 4$	We now have an equation with all the unknowns on one side. Subtract 3 from both sides so that all the constant terms are on the opposite side.	$5x = 4$
$5x = 4$ $\underline{\div 5 \div 5}$ $x = \dfrac{4}{5}$	Divide both sides by 5 to find the value of x. You can leave your answer as a fraction.	$x = \dfrac{4}{5}$

Exercise 2

1 Solve these equations.

a) $3x + 1 = 2x + 5$ b) $7x + 2 = 4x + 11$ c) $11x + 13 = 9x + 17$

d) $3x + 13 = 5x + 3$ e) $10x - 3 = 8x + 9$ f) $3x - 7 = x + 11$

g) $6x - 2 = 2x + 14$ h) $7x + 1 = 10x - 11$ i) $2x - 1 = 7x - 16$

2 Put these steps for solving the equation $2x + 7 = 19 - 2x$ in the right order.

$4x = 12$

$x = 3$

$4x + 7 = 19$

$-7 \qquad -7$

$+ 2x \qquad + 2x$

$\div 4 \qquad \div 4$

$2x + 7 = 19 - 2x$

> **Think about**
>
> How can you check to see if your solution to an equation is correct?

3 Solve these equations.

a) $2x + 4 = 13 - x$ b) $6x - 1 = 20 - x$ c) $24 - x = x + 4$ d) $60 - 2x = x + 3$

4 Solve these equations. Give any non-integer solutions as fractions.

a) $3t + 2 = t - 4$ b) $4x + 3 = 1 - 2x$ c) $5 - 2y = y + 11$ d) $1 + 4g = 2g + 2$

e) $5h - 2 = 3 - 5h$ f) $5 - 2b = 10 - b$ g) $7k - 17 = 7 - 5k$ h) $11p + 4 = 3p - 36$

▼ **Thinking and working mathematically activity**

Decide if these statements are true or false. Give a reason to explain your answers. Make up an equation for each true statement.

- The answer of an equation is always an integer.
- The answer of an equation could be a decimal number.
- The answer of an equation could be a fraction.
- The answer of an equation could be a negative number.

---

5 Solve these equations.

a) $2(4x + 1) = 7x + 13$ b) $3(x - 4) = x + 2$ c) $10(5x + 3) = 20x + 60$

d) $2(3x + 1) = 51 - x$ e) $3x + 14 = 2(5x - 21)$ f) $2x + 5 = 3(x - 1)$

6 Darlene is solving the equation $3(2x + 4) = 2(3x + 6)$

She writes:

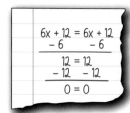

Is Darlene correct?

What does this mean about the equation? Is there a solution? Explain your answer.

7 Complete the equation below so that it has a solution of $x = 5$.

$2(3x + 7) = 10x \ldots$

8 Solve these equations.

a) $3x = 5(x - 2)$ b) $5(1 - 2m) = 9(1 - m)$ c) $5(2 - 3x) = -2(1 + 3x)$

d) $14 - 2(4 - r) = r$ e) $2(2v + 1) + v = 10v - 4$ f) $4(3x - 2) = 1 - (2x - 5)$

g) $8 + 0.8m = 2(0.75m - 3)$ h) $4(\frac{3}{2}t - 1) = 28 + 4t$

9 Each of the solutions below contains an error. For each one, describe the error and then solve the equation correctly.

a) $3(2x - 5) = 5 + x$
$3(2x) = 5 + x + 5$
$6x = x + 10$
$5x = 10$
$x = 2$

b) $8x - 3(3 - 2x) = -7$
$8x - 9 - 6x = -7$
$2x - 9 = -7$
$2x = 2$
$x = 1$

10 Mark and Cho have both started solving the equation $4(5x - 1) + 2 = 2(2x + 3)$.
Their first steps are shown below. Both are correct so far.

Mark:	Cho:
$4(5x - 1) + 2 = 2(2x + 3)$	$4(5x - 1) + 2 = 2(2x + 3)$
$20x - 4 + 2 = 4x + 6$	$2(5x - 1) + 1 = 2x + 3$

a) Describe Mark's first step.

b) Finish Mark's solution.

c) Describe Cho's first step.

d) Finish Cho's solution.

> **Discuss**
>
> Say which method you prefer, and why.

> **Think about**
>
> Without solving the equations, decide if it is always true, sometimes true or never true that:
>
> * $7(1 - 2n) = 7(34 - n)$ has the same answer as $13(1 - 2n) = 13(34 - n)$
> * In $7(1 - 2n) = 7(34 - n)$, you could multiply the brackets by any number. The answer will not change.

14.3 Constructing equations

Worked example 5

Ali and Ben each have the same number of pencils.

Ali has 4 full boxes of pencils and 2 loose pencils.

Ben has 2 full boxes of pencils and 10 loose pencils.

How many pencils are there in a full box?

Let x represent the number of pencils in a full box.	The unknown that we are trying to find is the number of pencils in a full box, so let's give this value a letter to represent it, x.	Visually, number of pencils in a full box x
Ali has $4x + 2$ pencils. Ben has $2x + 10$ pencils.	We can write an expression for the number of pencils that Ali and Ben have in terms of x.	Ali has 4 full boxes and 2 loose pencils. Ben has 2 full boxes and 10 loose pencils.
$4x + 2 = 2x + 10$	Since Ali and Ben have the same total number of pencils, we can make these two expressions equal.	Ali and Ben have the same number of pencils, so the overall bar lengths will be the same in our bar model. Ali: x | x | x | x | 2 Ben: x | x | 10

$4x + 2 = 2x + 10$
$-2x \quad\quad -2x$

We can now solve this equation to find the value of x.

$2x + 2 = 10$

Subtract $2x$ from both sides.

$2x + 2 = 10$
$\quad -2 \ -2$

$2x = 8$

Subtract 2 from both sides.

$2x = 8$
$\div 2 \ \div 2$

$x = 4$

Divide both sides by 2.

So the number of pencils in each box is 4.

$x = 4$

Worked example 6

One number is 6 more than another number.

The sum of twice the smaller number and three times the larger number is –7.
Construct an equation, and use it to find both numbers.

x $x + 6$ $2x + 3(x + 6) = -7$	Two numbers are mentioned in the equation. If you call the smaller number x, then the larger number is $x + 6$. 'Twice the smaller number' is $2x$. 'Three times the larger number' is $3(x + 6)$. The sum is –7. Now solve the equation.
$2x + 3x + 18 = -7$ $5x + 18 = -7$	First, expand the brackets on the left side.
$\quad -18 \ -18$ $5x = -25$	Subtract 18 from both sides.
$\quad \div 5 \ \div 5$ $x = -5$	Divide both sides by 5.
The two numbers are –5 and 1.	The smaller number, x, is –5. The larger number is $x + 6 = 1$.

Exercise 3

1 Multiplying a number by 3 and then adding 5 gives the same answer as adding 23 to the number.

What is the number?

2 Meera is thinking of a number, x. She doubles her number, adds 3 and then multiplies the result by 4. This gives her a final answer of 52.

a) Construct an equation in terms of x to represent this problem.

b) Solve your equation to find the value of x.

3 Angela is 12 years old. Her father's age is 11 less than 5 times Angela's age.

How old is her father?

4 Fiona and George each have the same number of eggs.

Fiona has 3 full boxes and 12 loose eggs.

George has 2 full boxes and 22 loose eggs.

How many eggs are there in a full box?

5 Quentin thinks of a number.

He multiplies it by 4 and then subtracts 21.

His answer is the same as the number he first thought of.

What number did Quentin think of?

6 Here is an isosceles triangle.

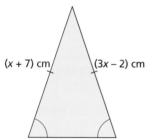

$(x + 7)$ cm $(3x - 2)$ cm

Find the value of x.

7 Find the sizes of the angles in the isosceles triangle below.

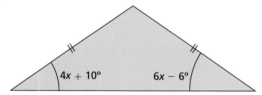

$4x + 10°$ $6x - 6°$

8 Here is a square.

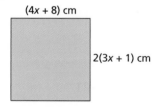

$(4x + 8)$ cm

$2(3x + 1)$ cm

a) Find the value of x.

b) Find the side length of the square.

9 By constructing and solving an equation, find the value of x.

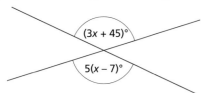

$(3x + 45)°$

$5(x - 7)°$

10 A triangle has sides of length x, $x + 2$ and $2x$. The perimeter of the triangle is 4 cm. Construct an equation and solve it to find x.

Thinking and working mathematically activity

A triangle has sides of length $x + 5$, $x + 10$ and $2x$.

- Can the perimeter of the triangle be a decimal number?
- Can the perimeter of the triangle be $3x$?
- Can the perimeter of the triangle be $6x$?
- Can the perimeter of the triangle be any positive number?

Give a reason for each answer.

---

11 By forming and solving an equation, find the value of the unknown in each diagram.

a)

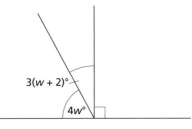

$3(w + 2)°$

$4w°$

b)

Diagrams not drawn accurately

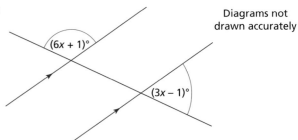

$(6x + 1)°$

$(3x - 1)°$

c)

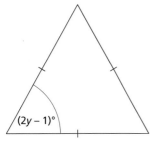

$(2y - 1)°$

d)

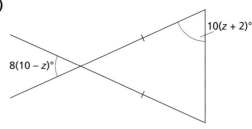

$8(10 - z)°$

$10(z + 2)°$

12 Sergey, Anish and Bethan each write an equation for the following problem:
'I think of a number, x. I subtract it from 6. Then I multiply the result by 3. I get −21.'

Sergey's equation is $3(x - 6) = -21$

Anish's equation is $3(6 - x) = -21$

Bethan's equation is $6 - 3x = -21$

Discuss whether each of these equations is correct or incorrect, and why. Then solve the correct equation to find x.

13 The diagram shows a regular pentagon and a square together with the lengths of one of their sides.

The pentagon has the same perimeter as the square.

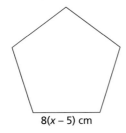

Diagrams not drawn accurately

3(x + 2) cm   8(x − 5) cm

Construct an equation, and solve it to find x.

14 **a)** The sum of two consecutive integers is 55.

What are the two integers?

b) The sum of two consecutive *even* integers is 110.

What are the two integers?

c) The sum of three consecutive *odd* integers is 51.

What are the three integers?

d) Create your own question similar to the ones above.

> **Tip**
>
> 'Consecutive' means following one another.
>
> 3 and 4 are consecutive numbers, and 12 and 14 are consecutive even numbers.
>
> Construct and solve an equation, rather than trying to guess possible answers.

14.4 Inequalities

Solving inequalities and inequalities on a number line

An open circle ○ is used to show: < (less than) and > (greater than)

A closed circle ● is used to show: ≤ (less than or equal to) and ≥ (greater than or equal to)

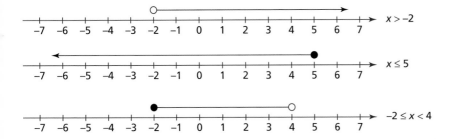

$x > -2$

$x \leq 5$

$-2 \leq x < 4$

> **Worked example 7**
>
> Joan is x years old and Ted is y years old. Write an inequality to show:
>
> **a)** Joan is younger than Ted.
>
> **b)** Their combined age is less than 35.
>
> **c)** Ted is no more than twice Joan's age.
>
> | **a)** $x < y$ | As Joan is younger than Ted, x is less than y. |
> |---|---|
> | **b)** $x + y < 35$ | Joan and Ted's ages added is less than 35. |
> | | x add y is less than 35. |
> | **c)** $y \leq 2x$ | Ted's age, y, is less than or equal to twice Joan's age ($2x$). |

1 I am thinking of a number, *n*. Write an inequality to show that:

  a) The number is between 1 and 10.

  b) Twice the number is less than or equal to 6.

  c) Double the number plus 3 gives an answer less than 8.

  d) Write down a value of *n* which satisfies all of these statements.

2 Write these inequalities in words.

  a) $x > 12$ **b)** $2 \leq x \leq 6$ **c)** $-1 < x < 6$

  d) $-3 \leq x$ **e)** $0 \leq x \leq 2$ **f)** $-2 < x \leq 2$

  g) Show the inequalities on a number line.

3 Write down the inequality that each number line represents.

  a)
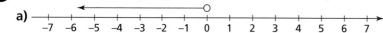

  b)
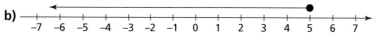

  c)
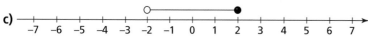

  d)
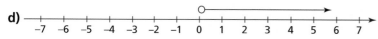

  e)
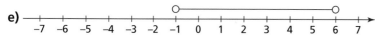

4 **a)** Write down the smallest whole number that satisfies the inequality $x \geq -5$.

  b) Write down the largest whole number that satisfies the inequality $x < -3$.

5 Write down the integers that satisfy these inequalities and show the inequality on a number line.

  a) $0 < x < 4$ **b)** $-3 \leq x < 1$ **c)** $-1 < x \leq 5$

6 Danika is *d* years old and Ellen is *e* years old.

  Write an inequality to show:

  a) Danika is younger than 13

  b) Danika is at least 3 years older than Ellen

  c) Ellen is at least 6 years old

  d) Write this inequality in words: $d - e \leq 4$.

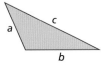

Thinking and working mathematically activity

Determine if it is always true, sometimes true or never true that, in a triangle with sides a, b and c:

- $a + b = c$
- $a + b < c$
- $a + b > c$

Draw different triangles to investigate. Include right angle triangles, isosceles triangles and equilateral triangles.

7 Find the inequalities that are equivalent to $x > 4$:

a) $2x > 8$ b) $x - 2 > 4 - 2$ c) $x - 4 < 0$

d) $x + 3 > 1$ e) $5x > 20$ f) $3x > 7$

8 Find the odd one out. Give a reason for your answer.

a) $r \leq 3$ b) $4r \leq 12$ c) $r + 4 \leq 6$ d) $9r \leq 27$

9 If $y > 5$, copy and complete each of these inequalities.

a) $y - 3 > \ldots\ldots$ b) $4y > \ldots\ldots$ c) $2y + 1 > \ldots\ldots$

Thinking and working mathematically activity

Write four different inequalities that are equivalent to:

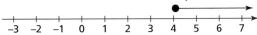

Explain why they are equivalent to the inequality shown.

Consolidation exercise

1 Solve the equations.

a) $5(2x - 1) = 35$ b) $\frac{5}{6}y = 45$ c) $3x + 1 = 6x + 19$

d) $5n - 2 = 1 - 3n$ e) $\frac{2}{3}p + 7 = 15$

2 Solve the equations.

a) $5 - 3b = 2 - b$ b) $4x = 12(2 - x)$ c) $5(1 - 2m) = 9(1 - m)$

d) $\frac{7}{2} - y = 2$ e) $9 - 3(4 - r) = r$ f) $2(2v + 5) + v = 7v - 4$

3 Sam has solved the three equations below.

Use substitution to check whether each solution is correct.
Write a correct solution for any that are incorrect.

a) $3(1 - b) = 5 - 2b$ Sam's solution: $b = 2$

b) $2k - 1 = 3 - (1 - k)$ Sam's solution: $k = 3$

c) $5(z - 3) = 3(z - 5)$ Sam's solution: $z = 15$

4 Here is an equation with a missing term: $3(5 - 4x) = \boxed{} - 6x$.

If the missing term is 27, the equation is $3(5 - 4x) = 27 - 6x$, with solution $x = -2$.

Find the value of the missing term if the solution is:

a) 0   b) −5   c) 1

5 Omid says,

'I think of a number, x. I multiply it by 3. Then I add 5. I get −13.'

Construct an equation and solve it to find.

6 The diagram shows a kite.

Find the value of x.

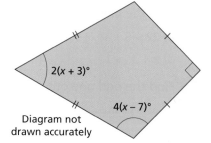

2(x + 3)°

4(x − 7)°

Diagram not drawn accurately

7 Ji-yoo is trying to solve the equation $3(x + 3) = 9$.

She says, 'My first step is to divide both sides by 3, so I get the equivalent equation $x + 1 = 3$. The solution is $x = 2$.'

a) Explain why this is incorrect.

b) Solve the equation using a different first step.

c) It is possible to solve the equation by starting with Ji-yoo's first step, 'divide both sides by 3'?
Solve the equation by starting with this step.

8 Insert the correct inequality sign in these statements.

a) If $n \leq 6$, then $2n$ 12   b) If $n < 5$, then $n + 4$ 9

c) If $n > -2$, then $8 - n$ 10   d) If $n \leq -2$, then n^3 8

End of chapter reflection

You should know that...	You should be able to...	Such as...
You must do the same to both sides of an equation to keep it balanced. You can check whether your solution is correct by substituting it back into the equation and making sure the two sides are equal.	Solve an equation with an unknown on one or both sides with and without brackets.	Solve: a) $\frac{3}{4}x - 2 = 25$ b) $2(5x + 6) = 102$ c) $2x + 31 = 7 - 2x$
Equations can be constructed to help solve problems.	Construct an equation to represent a problem and interpret the solution in the context of the problem.	A triangle has angles of x, $2x - 40°$ and $90° - x$. Find the value of the smallest angle.
Inequalities can be shown on a number line. An open circle is used for < or >. A closed circle is used for ≤ or ≥.	Draw an inequality on a number line or write an inequality from a number line.	Draw a number line to show these inequalities a) $x > 4$ b) $-2 \leq x < 6$ Write down the inequality shown on this number line.

You will learn how to:

- Use knowledge of coordinates to find the midpoint of a line segment.

Starting point

Do you remember...

- how to write the coordinates of a point on a grid?

  For example, write the coordinates of each of the lettered points on this grid.

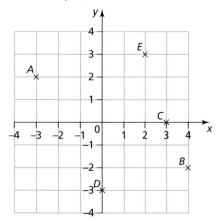

- how to add and subtract negative integers?

  For example, find the difference between 9 and –3

- how to find the mean value of two numbers?

  For example, find the mean of 8 and –2

This will also be helpful when...

- you find the perpendicular bisector of a line segment.

15.0 Getting started

Here is a triangle *ABC*.
Copy the triangle onto squared paper.

- Use a ruler to mark the points *P, Q* and *R* onto your diagram where:
  P is on the side *AB*, halfway between *A* and *B*
  Q is on the side *BC*, halfway between *B* and *C*
  R is on the side *AC*, halfway between *A* and *C*.

- Draw the lines *AQ* and *BR*.
  These lines are called **medians** of the triangle.
  The point where they meet is called the **centroid**.
  What are the coordinates of the centroid?

- Draw the third median line *CP*.

- What do you notice about all three median lines?

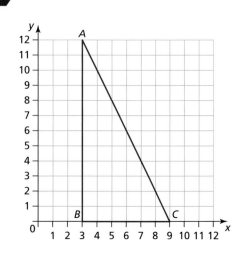

15.1 Midpoint of a line segment

Key terms

A **line segment** is a section of a line. It has two end points.

The **midpoint** of a line segment is the point on it that is the same distance from both end points. It is halfway along the line segment.

▼ Thinking and working mathematically activity

Draw a pair of axes on squared paper and draw several line segments. The end points of each line segment should have integer coefficients.

- Find the midpoint of each of your line segments.
- Investigate how the coordinates of each midpoint relate to the coordinates of the ends of the line segment.
- Write a rule for finding the coordinates of the midpoint of any line segment.

Worked example 1

A and B are the points with coordinates (–2, 5) and (4, 2).

Find the coordinates of the midpoint of the line segment AB.

$$\frac{-2 + 4}{2} = \frac{2}{2} = 1$$

$$\frac{5 + 2}{2} = \frac{7}{2} = 3.5$$

So the coordinates of the midpoint are (1, 3.5)

Find the mean of the x-coordinates of A and B.

Then find the mean of the y-coordinates.

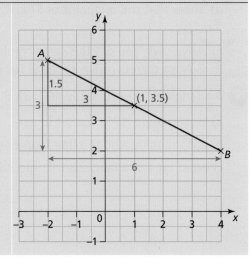

Exercise 1

1 Write down the coordinates of the midpoint of the line segment joining each pair of points.

a) *A* and *B*

b) *A* and *C*

c) *B* and *C*

d) *B* and *E*

e) *A* and *D*

f) *C* and *E*

g) *E* and *F*

h) *B* and *G*

i) *E* and *G*

j) *A* and *G*

2 Find the midpoint of the line segment joining each pair of points.

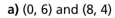

a) (0, 6) and (8, 4)   **b)** (9, 2) and (1, 12)   **c)** (7, 5) and (2, 13)

d) (3, 2) and (10, 7)   **e)** (−1, 6) and (9, −10)   **f)** (9, −4) and (12, 16)

g) (13, 5) and (31, 17)   **h)** (4, −2) and (−5, −6)   **i)** (13, 25) and (−7, −7)

3 *ABCD* is a kite with coordinates *A*(3, 7), *B*(7, 10), *C*(13, 2) and *D*(3, 2).

a) Find the coordinates of the midpoint of the diagonal *AC*.

b) Find the coordinates of the midpoint of the diagonal *BD*.

4 *P* has coordinates (11, 17). *Q* has coordinates (3, 1). *R* is the midpoint of the line segment *PQ*.

a) Find the coordinates of *R*.

b) Find the coordinates of *S*, the midpoint of *PR*.

c) Find the coordinates of *T*, the midpoint of *RQ*.

d) Show that the midpoint of *ST* is *R*.

e) Anya says 'You get this same result for different coordinates of *P* and *Q*.'
Is Anya correct? Give a reason for your answer.

5 A triangle has coordinates *A*(−1, 3), *B*(5, 11), and *C*(5, 3).

a) Plot triangle *ABC* on a set of axes.

b) *C* is the midpoint of a line segment *AD*. Find the coordinates of *D*.

c) What type of triangle is triangle *ABD*?

6 Find the missing coordinates from each row in this table.

	A	B	Midpoint of AB
a)		(7, 8)	(6, 5)
b)		(−4, 23)	(0, 18)
c)	(8, 3)		(4, 4.5)
d)		(7, 11)	(5, 7)
e)	(−3, 17)		(1, 22)
f)	(6, −5)		(8.5, 1)

> **Tip**
>
> In question 6, sketch a diagram and think about the differences in the *x*-coordinates and the differences in the *y*-coordinates.

7 The points *L* and *M* have coordinates *L*(*a*, *c*) and *M*(*b*, *d*).

The values of *a*, *b*, *c* and *d* are all square numbers with
   a < *b* and *c* < *d*.

The coordinates of the midpoint of *LM* are (20, 25).

Find the values of *a*, *b*, *c* and *d*.

8 *ABCD* is a parallelogram where *A* is (−3, −7) and *B* is (−5, −1).

The coordinates of the midpoint of each diagonal are (0, 0).

Find the midpoint of *CD*.

> **Discuss**
>
> If the midpoint of two points is (0, 0), How are the coordinates of the two points related?

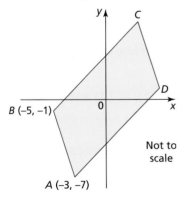

Not to scale

9 *A* and *B* are points with coordinates (3*d*, −*d*) and (−5*d*, 5*d*).
   C is the midpoint of *AB*.

a) Find the coordinates of *C* in terms of *d*.

b) The coordinates of *C* are (−6, 12). What are the coordinates of *A*?

Thinking and working mathematically activity

Technology question Use dynamic geometry software to draw any triangle *ABC*.

- Find the midpoints of each side.
- The lines connecting the midpoints divide triangle *ABC* into four triangles. How are these triangles related?
- Investigate for a range of different starting triangles.

Consolidation exercise

1 Here are some coordinates of points.

> (17, 4) (10, 11) (11, 9) (6, 6) (13, 10) (14, 10)
> (10, 5) (9, 7) (3, 12)

Choose from the box four pairs of points that have a midpoint at (10, 8).

2 Here are four statements.

| **Statement A** |
| The midpoint of the line segment joining (4, 10) and (10, 12) is (6, 2) |

| **Statement B** |
| The midpoint of the line segment joining (7, 3) and (7, –3) is on the x-axis. |

| **Statement C** |
| The midpoint of the line joining (2, 10) and (10, 2) is the same as the midpoint of the line joining (2, 2) and (10, 10) |

| **Statement D** |
| *O* is the point (0, 0) and *P* is the point (0, 24). If *Q* is the midpoint of *OP*, then the midpoint of *OQ* is closer to *P* than it is to *O*. |

Which of the statements are true and which are false?

3 If the midpoint of two points, *A* and *B*, in on the y-axis, then both *A* and *B* must be on the y-axis.

Give an example to show that Mica's statement is false.

4 *A* has coordinates (35, ■). *B* has coordinates (■, ▲). The midpoint of the line segment *AB* is (43, 41). Find the values of ■ and ▲.

5 The midpoints of the four sides of a quadrilateral are at the points (2, 1), (6, –1) and (0, –3).

Find possible coordinates of the 4 vertices of the quadrilateral.

6 **Vocabulary question** Use the words from the box to copy and complete the sentence below.

| add | line segment | mean |
| midpoint | halfway | endpoints |

A section of a line joining two points *A* and *B* is called a _____ . The point on this line that is _____ between the two endpoints is called the _____ To find the x-coordinate of the midpoint, _____ together the x-coordinates of *A* and *B* and then divide by 2. Similarly the y-coordinate of the midpoint is the _____ of the y-coordinate of the _____ .

End of chapter reflection

You should know that...	You should be able to...	Such as...
The midpoint of a line segment is the point on it that is halfway between the two points. The coordinates of the midpoint can be found from the mean of the coordinates of the endpoints.	Find the coordinates of the midpoint of a line segment.	A is the point (8, 5) B is the point (14, 3) Find the coordinates of the midpoint of AB.

16 Fractions, decimals and percentages

You will learn how to:
- Recognise fractions that are equivalent to recurring decimals.
- Understand the relative size of quantities to compare and order decimals and fractions (positive and negative), using the symbols =, ≠, >, <, ≤ and ≥.

Starting point

Do you remember...

- how to convert between terminating decimals, fractions and percentages?
  For example, write $\frac{3}{25}$ as a decimal and as a percentage.
- how to compare and order positive decimals and fractions, using the symbols =, ≠, > and <?
  For example, write three correct statements about $\frac{2}{5}$, 0.38 and $\frac{9}{20}$.
  Use one of the symbols =, ≠, > or < in each statement.

This will also be helpful when...

- you learn to see whether a fraction has a recurring or terminating decimal equivalent (without doing a division calculation).

16.0 Getting started

When you write any fraction between 0 and 1 as a decimal, you make the decimal from tenths, hundredths, thousandths, etc.

How would you write $\frac{1}{3}$ as a decimal?

Use a strip of paper to represent 1. By first cutting it into tenths, then hundredths, then thousandths, etc., as needed, try to divide it into three equal piles, so that each pile represents $\frac{1}{3}$ as a decimal.

Suggest how to write $\frac{1}{3}$ as a decimal. Explain your answer.

16.1 Recurring decimals

Key terms

A **terminating decimal** is a decimal that stops after a number of decimal places, for example 0.6 or 0.74 or 0.31406

A **recurring decimal** has digits that continue forever in a repeating pattern.
We write dots or a bar above the part that repeats.

$0.\dot{2}$ or $0.\bar{2}$ means that the 2 repeats: 0.2222...

Two dots or a bar above show the beginning and end of a recurring group of numbers.

$0.\dot{1}2\dot{3}$ or $0.\overline{123}$ means that the 123 repeats: 0.123123123...

Worked example 1

Express each fraction as a recurring decimal.

a) $\dfrac{7}{9}$

b) $\dfrac{15}{22}$

a) $\dfrac{7}{9} = 7 \div 9$ $$\begin{array}{r} 0\ .\ 7\ 7\ ... \\ 9\overline{)\ 7\ .\ {}^{7}0\,{}^{7}0\,{}^{7}0} \end{array}$$ $\dfrac{7}{9} = 0.\dot{7}$	To convert to a decimal, divide the numerator by the denominator. When the remainder is 7, you then find $70 \div 9 = 7$, which has remainder 7. This pattern will continue forever.	
b) $\dfrac{15}{22} = 15 \div 22$ $$\begin{array}{r} 0\ .\ 6\ 8\ 1\ 8\ 1\ ... \\ 22\overline{)\ 15\ .\ {}^{15}0\,{}^{18}0\ {}^{4}0\,{}^{18}0\ {}^{4}0} \end{array}$$ $\dfrac{15}{22} = 0.6\dot{8}\dot{1}$	When the remainder is 18, you then find $180 \div 22 = 8$, which has remainder 4. When the remainder is 4, you then find $40 \div 22 = 1$, which has remainder 18. This pattern will continue forever.	You can also write this as a long division (but it is harder to see the remainders): $$\begin{array}{r} 0.6\ 8\ 1\ 8\ 1\ ... \\ 22\overline{)\ 15.0\ 0\ 0\ 0\ 0} \\ \underline{13\ \ 2} \\ 1\ 8\ 0 \\ \underline{1\ 7\ 6} \\ 4\ 0 \\ \underline{2\ 2} \\ 1\ 8\ 0 \\ \underline{1\ 7\ 6} \\ 4\ 0 \end{array}$$

Think about

How can you see when a decimal will recur? Use the solution to part **b)** of Worked example 1 to
explain how you can recognise this from the division calculation.

Exercise 1   1 - 2, 4 - 6, 8

1 Rewrite each recurring decimal using dot notation.

   a) 0.444...   b) 2.3535...   c) 0.8333...   d) 57.0481481...

2 Use a written division method to convert these fractions to recurring decimals.

   Write the recurring decimals using dot notation.

   a) $\dfrac{1}{3}$ b) $\dfrac{1}{6}$ c) $\dfrac{1}{9}$ d) $\dfrac{1}{12}$ e) $\dfrac{6}{11}$ f) $\dfrac{7}{18}$

   g) $\dfrac{2}{15}$ h) $\dfrac{4}{37}$ i) $\dfrac{5}{6}$ j) $\dfrac{31}{44}$ k) $\dfrac{9}{11}$ l) $\dfrac{19}{27}$

3 Use a calculator to convert these fractions to recurring decimals.

Write the recurring decimals using dot notation.

a) $\dfrac{31}{54}$ b) $\dfrac{9}{88}$ c) $\dfrac{17}{74}$ d) $\dfrac{30}{41}$

4 Maya says, 'Any fraction less than 1 that has a denominator of 6 gives a recurring decimal.'

Is Maya correct? Explain your answer.

> **Tip**
>
> Look for an example that does not fit Maya's rule.

5 Round these recurring decimals to 3 decimal places.

a) 0.5555...   b) $6.1\dot{2}\dot{8}$ c) $0.78\dot{3}$ d) $0.\dot{8}$

6 Mtambwa wants to convert $\dfrac{9}{40}$ to a decimal.

His working is below.

$$40 \overline{)\,9.\,^90\,^{10}0\,}^{\;0.2\;2\;...}$$

$$\dfrac{9}{40} = 0.\dot{2}$$

a) Describe the mistake Mtambwa has made.

b) Find the correct answer.

7 **Technology question** Using a spreadsheet program, create a table showing the decimal

equivalents of sevenths, from $\dfrac{1}{7}$ to $\dfrac{6}{7}$. Show at least nine decimal places for each decimal.

Describe any pattern you can see.

8 a) Use a written division method to find the decimal equivalents of ninths from $\dfrac{1}{9}$ to $\dfrac{4}{9}$.

b) Describe any pattern you find. Use it to predict the decimal equivalent of $\dfrac{5}{9}$ and $\dfrac{8}{9}$.

Check your prediction by finding $\dfrac{5}{9}$ and $\dfrac{8}{9}$ using written division.

c) Suggest an explanation for the pattern.

> **Discuss**
>
> Does $0.\dot{9} = 1$?

Thinking and working mathematically activity

Technology question Using a spreadsheet program, find the decimal equivalents of fractions with numerator 1 and denominators from 2 to 100.

List the fractions that have one repeating digit. Look for any patterns in their denominators.

Then look for patterns for fractions with two, three and five repeating digits.

What do you notice about fractions with four repeating digits?

Key terms

The sign ≤ means '**less than or equal to**'. Examples are: 3 ≤ 3 and 3 ≤ 99.

The sign ≥ means '**greater than or equal to**'. Examples are: 17 ≥ 17 and 17 ≥ –2.

The statement	The statement
4.1 < x < 8.6	4.1 ≤ x ≤ 8.6
tells you that x lies between 4.1 and 8.6. x cannot equal 4.1 or 8.6.	tells you that x can be between 4.1 and 8.6, but it can also equal 4.1 or 8.6.
The range of x is 4.1 **exclusive** to 8.6 exclusive.	The range of x is 4.1 **inclusive** to 8.6 inclusive.
(This means not including 4.1 and not including 8.6)	(This means including 4.1 and including 8.6)

Worked example 2

a) If –1.89 ≤ x ≤ –0.61, write the greatest and least possible values of x.

b) If –0.56 ≤ x < 0.1, which numbers from the list below are possible values of x?

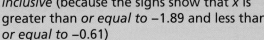

| 0.1 | –0.55 | –0.56 | 0.56 | 0 | –0.06 |

a) The greatest possible value of x is –0.61 The least possible value of x is –1.89	The range of x is –1.86 *inclusive* to –0.61 *inclusive* (because the signs show that x is greater than *or equal to* –1.89 and less than *or equal to* –0.61)	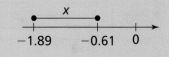
b) The possible values of x are: –0.55, –0.56, 0 and –0.06	The range of x is –0.56 *inclusive* to 0.1 *exclusive* (because the signs show that x is greater than *or equal to* –0.56 but less than 0.1) So x can be –0.56 but cannot be 0.1. It can also be any number between –0.56 and 0.1	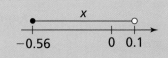

Worked example 3

a) Compare $\frac{3}{4}$ and $\frac{3}{5}$ without changing the denominators or converting to decimals or percentages.

Write a statement using <, > or =

b) Compare $\frac{3}{5}$ and $\frac{9}{14}$ by making the numerators the same. Write a statement using <, > or =

a) $\frac{3}{4} > \frac{3}{5}$	Both numerators are 3, so compare the denominators. $\frac{3}{4}$ is greater than $\frac{3}{5}$ because quarters are bigger than fifths.	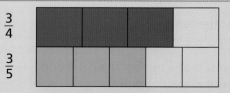
b) $\frac{3}{5} = \frac{9}{15}$ $\frac{9}{15} < \frac{9}{14}$ So, $\frac{3}{5} < \frac{9}{14}$	Multiply the numerator and denominator of $\frac{3}{5}$ by 3. Now both numerators are 9. $\frac{9}{15}$ is less than $\frac{9}{14}$ because fifteenths are smaller than fourteenths.	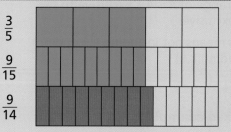

Worked example 4

a) Write these numbers in order from lowest to highest.

$-1\frac{2}{9}$ $\qquad$ $\frac{2}{9}$ $\qquad$ $\frac{2}{3}$ $\qquad$ $-\frac{5}{6}$ $\qquad$ $\frac{5}{6}$

b) Write these numbers in order from lowest to highest.

0.401 $\qquad$ −0.444 $\qquad$ −0.45 $\qquad$ −0.409 $\qquad$ 0.41

c) Write these numbers in order from lowest to highest.

$-\frac{4}{5}$ $\qquad$ $-\frac{21}{25}$ $\qquad$ −0.799 $\qquad$ $-\frac{7}{8}$ $\qquad$ −0.85

a) $-1\frac{4}{18}, \frac{4}{18}, \frac{12}{18}, -\frac{15}{18}, \frac{15}{18}$ From lowest to highest: $-1\frac{4}{18}, -\frac{15}{18}, \frac{4}{18}, \frac{12}{18}, \frac{15}{18}$ or $-1\frac{2}{9}, -\frac{5}{6}, \frac{2}{9}, \frac{2}{3}, \frac{5}{6}$	Make the denominators the same. Then order the fractions, taking care with signs.	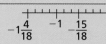
b) From lowest to highest: −0.45, −0.444, −0.409, 0.401, 0.41		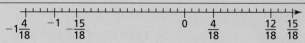

c)

$-0.8, -0.84, -0.799, -0.875, -0.85$

From lowest to highest:

$-0.875, -0.85, -0.84, -0.8, -0.799$

or

$-\dfrac{7}{8}, -0.85, -\dfrac{21}{25}, -\dfrac{4}{5}, -0.799$

Convert the fractions to decimals:

$-\dfrac{4^{\times 20}}{5_{\times 20}} = -\dfrac{80}{100} = -0.8$

$-\dfrac{21^{\times 4}}{25_{\times 4}} = -\dfrac{84}{100} = -0.84$

$-\dfrac{7^{\times 125}}{8_{\times 125}} = -\dfrac{875}{1000} = -0.875$

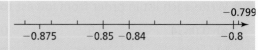

Exercise 2

1 If $-1.7 \leq x \leq 2.5$, write:

   a) the lowest possible value of x

   b) the highest possible value of x

   c) all possible integer values of x.

> **Tip**
>
> Sketching number lines can help you with the questions in this exercise.

2 Write each statement using mathematical symbols.

   a) The range of x is 6 inclusive to 22 exclusive.

   b) n is greater than 0.18 and less than or equal to 0.98

   c) The range of x is $\dfrac{1}{3}$ inclusive to $\dfrac{4}{5}$ inclusive.

   d) The range of p is -5 exclusive to -0.42 exclusive.

3 Write < or > to make correct statements.

   a) $\dfrac{5}{11}$ _____ $\dfrac{6}{11}$ **b)** $\dfrac{5}{11}$ _____ $\dfrac{5}{12}$ **c)** $-\dfrac{3}{7}$ _____ $-\dfrac{4}{7}$ **d)** $-\dfrac{3}{4}$ _____ $-\dfrac{3}{5}$

4 In each pair, rewrite the fractions with the same denominator. Then write < or > to make correct statements.

   a) $\dfrac{3}{8}$ _____ $\dfrac{5}{12}$ **b)** $-\dfrac{2}{3}$ _____ $-\dfrac{4}{9}$ **c)** $-\dfrac{3}{10}$ _____ $-\dfrac{4}{15}$ **d)** $-\dfrac{2}{5}$ _____ $-\dfrac{3}{7}$

5 In each pair, rewrite the fractions with the same numerator. Then write < or > to make correct statements.

   a) $\dfrac{10}{13}$ _____ $\dfrac{5}{7}$ **b)** $\dfrac{3}{8}$ _____ $\dfrac{9}{25}$ **c)** $\dfrac{5}{12}$ _____ $\dfrac{3}{7}$ **d)** $-\dfrac{3}{11}$ _____ $-\dfrac{2}{9}$

6 Aino wants to find which is greater, $\dfrac{4}{29}$ or $-\dfrac{7}{49}$. Her working is below.

$\dfrac{4}{29} = \dfrac{28}{203}$ and $-\dfrac{7}{49} = -\dfrac{28}{196}$

196 is less than 203, so $-\dfrac{28}{196}$ is greater than $\dfrac{28}{203}$

$-\dfrac{7}{49} > \dfrac{4}{29}$

Is she correct? Explain your answer.

7 Write each list in order from lowest to highest.

    a) 0.0854   −0.9   0.34   −0.134   −0.95

    b) −1.96   −1.906   −0.98   −1.955   −0.906

    c) $-2\dfrac{1}{3}$ $-\dfrac{3}{2}$ $-2\dfrac{1}{4}$ $-1\dfrac{5}{6}$ $-\dfrac{13}{6}$

    d) −0.85 $-\dfrac{22}{25}$ −0.818 $-\dfrac{4}{5}$ −0.901

8 Write < or > to make correct statements.

    a) 0.581 _____ 0.59   **b)** −0.86 _____ −0.861   **c)** −0.7 _____ −0.67   **d)** −0.1 _____ 0.03

9 Write = or ≠ to make correct statements.

    a) $4\dfrac{7}{20}$ _____ 4.45   **b)** $-\dfrac{2}{5}$ _____ −0.2   **c)** $\dfrac{1}{3}$ _____ 0.3   **d)** 1.125 _____ $\dfrac{9}{8}$

10 If $-0.86 < x \le -0.22$, find the numbers from the list below that are possible values of x.

    −0.2   0   −0.861   −0.859   −0.202   −0.86   −0.22

11 If $-\dfrac{1}{2} \le x < -\dfrac{1}{6}$, find the numbers from the list below that are possible values of x.

    $-\dfrac{1}{3}$ $-\dfrac{1}{2}$ $-\dfrac{3}{4}$ $-\dfrac{1}{4}$ $-\dfrac{1}{6}$

12 Decide whether each statement is true or false.

    a) $-3.04 < -3.4$ **b)** $\dfrac{2}{3} \ge 0.\dot{6}$ **c)** $0.85 \le \dfrac{17}{20}$ **d)** $\dfrac{5}{9} \ne 0.\dot{5}$

▼ Thinking and working mathematically activity

Where possible, find a number that appears in both intervals.

If there is no number in both intervals, give a reason why.

- $-2 \le x < 3.5$ and $-4.6 < x \le -1.3$

- $-1.55 \le x \le -0.4$ and $-\dfrac{2}{5} < x \le 0.15$

- $\dfrac{1}{12} < x < \dfrac{1}{3}$ and $\dfrac{1}{4} < x \le \dfrac{1}{2}$

- $\dfrac{7}{100} < x \le \dfrac{11}{25}$ and $0.44 \le x < 0.5$

- $-0.56 < x \le -\dfrac{1}{3}$ and $-0.3 \le x < 0$

- $0 < x < 0.01$ and $\dfrac{1}{200} < x < 0.1$

16.3 Comparing fractions by converting to percentages

Worked example 5

Which school has the highest proportion of teachers per student?

School	Students	Teachers
A	750	37
B	800	47
C	710	36

School A

37 as a percentage of 750 is

$\frac{37}{750} \times 100\% = 4.9\dot{3}\%$

School B

47 as a percentage of 800 is

$\frac{47}{800} \times 100\% = 5.875\%$

School C

36 as a percentage of 710 is

$\frac{36}{710} \times 100\% = 5.07042...\%$

School B has the highest proportion of teachers per student.

For each school, write the proportion of teachers per student as a fraction.

Then use a calculator to convert the fraction to a percentage.

Another way to write the first calculation is

$37 \div 750 \times 100\%$

Did you know?

The brain has about 2% of a person's mass but it uses about 20% of their energy.

Exercise 3

1, 3–7

1. Write each of the following as a fraction and as a percentage. Write the fractions in their simplest form.

   a) 18 out of 20
   b) 36 minutes out of 1 hour
   c) $36 out of $45
   d) 12 out of 150
   e) 14 out of 40
   f) 78 kg out of 120 kg

2. Express each of the following as a percentage. Round each percentage to 1 decimal place.

   a) 37 bikes out of 149 bikes
   b) 130 apples out of 635 apples
   c) 3 boys out of 295 boys
   d) 67 cups out of 201 cups
   e) 80 minutes out of 2 hours
   f) 88 cm out of 3.44 m

3 Write <, > or = to complete each statement.

a) 59% _____ $\frac{3}{5}$ b) –44% _____ $-\frac{11}{20}$ c) $\frac{9}{25}$ _____ 36.6%   d) 277% _____ $\frac{11}{4}$

4 Jake wants to convert $\frac{3}{8}$ to a percentage. His working is below.

If $\frac{1}{4}$ is 25%, then $\frac{1}{8}$ is 50% and $\frac{3}{8}$ is 150%.

Amy thinks this is incorrect because $\frac{3}{8}$ is less than 1, so it should be less than 100%.

Who is correct? Find the correct percentage equivalent of $\frac{3}{8}$.

5 Write each set in order, starting with the lowest.

a) 0.88 $\frac{17}{20}$ 81% $\frac{41}{50}$ 39 out of 50

b) 0.6 $\frac{2}{3}$ 61%   0.605   66.5%

c) $-\frac{1}{4}$ –0.24   –24.9% $-\frac{1}{5}$ –0.251

d) 11% $-\frac{3}{25}$ 0.1   –0.199 $\frac{1}{9}$

6 If ● and ■ are different integers less than 10 and $\frac{●}{■} \geq 85\%$, what could the values of ● and ■ be?

Can you find more than one answer?

7 These are Tom's results in exams in three subjects.

Maths	English	Science
$\frac{139}{160}$	82.5%	40 out of 50

In which subject did he achieve the best result? Show your reasoning.

8 Jade compares the amount of sugar in three different type of fruit juice.

- Pineapple: 23.8 g of sugar in 250 ml
- Orange: 25.5 g of sugar in 300 ml
- Blackcurrant: 22.7 g of sugar in 200 ml

Jade says that orange juice has the highest proportion of sugar.
Is she correct? Show how you worked out your answer.

9 The table shows the number of children and adults who visited a zoo one weekend.

	Total number of people	Children
Saturday	3610	2300
Sunday	3400	1978

Showing your calculations clearly, find which day had the larger proportion of children visiting.

Thinking and working mathematically activity

Write four **different** digits (not including zero) in the boxes to make a positive fraction (for example, $\frac{23}{61}$).

Find the equivalent percentage. You can use a calculator.

Try to make:

* the smallest possible percentage that is less than 100%
* the largest possible percentage that is less than 100%
* the smallest possible percentage that is greater than 100%
* the largest possible percentage that is greater than 100%.

Discuss your methods. Explain how you know your answers are correct.

Consolidation exercise

1 Match each fraction with its decimal equivalent.

$\frac{19}{21}$	$0.7\overset{\bullet}{9}1\overset{\bullet}{6}$
$\frac{19}{22}$	$0.\overset{\bullet}{9}0476\overset{\bullet}{1}$
$\frac{19}{24}$	$0.730769230\overset{\bullet}{}$
$\frac{19}{25}$	$0.\overset{\bullet}{8}6\overset{\bullet}{3}$
$\frac{19}{26}$	0.76

> **Tip**
>
> Think about how to match these without using division.

2 Jamie wants to answer the question, 'Find $2 \div 9$ and write the result correct to 2 decimal places.'

He says, 'The answer is $0.2\overset{\bullet}{2}$.'

Is Jamie correct? Explain your answer.

3 Find the missing divisor in each calculation.

a) $\begin{array}{r} 0.1\ 3\ 3\ldots \\ ? \overline{)\ 2.\,{}^2 0\ {}^5 0\ {}^5 0} \end{array}$

b) $\begin{array}{r} 0.2\ 0\ 8\ 3\ 3\ldots \\ ? \overline{)\ 5.0\,{}^2 0\,{}^{20} 0\,{}^8 0\,{}^8 0} \end{array}$

c) Write each of the recurring decimals above using dot notation.

4 Decide whether each statement is always true, sometimes true or never true.
Explain your answers.

    a) A fraction with a larger denominator is smaller than a fraction with a smaller denominator.

    b) A fraction with a larger numerator is larger than a fraction with a smaller numerator.

5 A cinema records the types of snack that its customers buy.

    $\frac{3}{8}$ of the customers buy popcorn.

    $\frac{7}{20}$ of the customers buy chocolate.

    $\frac{3}{5}$ of the customers buy nachos.

    a) Which snack is bought by the most customers? Show your reasoning.

    b) Which snack is bought by the fewest customers? Show your reasoning.

6 In each statement there is a decimal missing. Write down a possible value for each.

    a) $-0.7 > \underline{\hspace{1cm}} > -0.9$ **b)** $-0.02 < \underline{\hspace{1cm}} \leq 0.01$ **c)** $-9.34 < \underline{\hspace{1cm}} < -9.33$

7 Luka says, '$\frac{1}{3}$ is the same as 33%.'

    Is he correct? Explain how you know.

8 A cake recipe uses 140 g of butter out of a total of 400 g of ingredients.

    Another cake recipe uses 90 g of butter out of a total of 250 g of ingredients.

    Show which recipe uses the higher percentage of butter.

9 Last year (which was not a leap year), the weather was dry for 219 days.
What percentage of days were wet?

10 Maryann scored 23/25 in a Music test, 11/15 in a History test and 67/80 in a Geography test.

    a) Express each mark as a percentage.

    b) In which subject did she get her best score?

11 Class A has 25 students and Class B has 20 students.

    One day there are 19 students in Class A and 17 students in Class B.

    Class A's teacher says, 'Our attendance is better because we have
two more students in school today than Class B'.

    Do you agree with this? Explain your answer.

End of chapter reflection

You should know that...	You should be able to...	Such as...
A recurring decimal has digits that continue forever in a repeating pattern.	Write a recurring decimal using dot notation.	Write 0.5333... using dot notation. Use a calculator to convert $\frac{10}{11}$ to a recurring decimal. Write the decimal using dot notation.
Some fractions have equivalent decimals that are recurring.	Convert a fraction to a recurring decimal using written division.	Use written division to convert $\frac{5}{18}$ to a recurring decimal.
The symbol ≤ means 'less than or equal to'. The symbol ≥ means 'greater than or equal to'.	Understand and use the symbols ≤ and ≥.	Decide whether each statement is true or false. $\frac{4}{9} \geq 0.4$ $\frac{1}{8} \leq 0.125$
If x is to the left of y on the number line, then: • x is less or lower than y, and • y is greater than or higher than x.	Compare and order positive and negative fractions and decimals.	Write these decimals in order from lowest to highest. −0.25 −0.251 −0.2 −0.202 Write these fractions in order from lowest to highest. $-\frac{5}{4}$ $-1\frac{1}{8}$ $-\frac{17}{12}$ $-1\frac{1}{6}$

17 Presenting and interpreting data 2

You will learn how to:

- Record, organise and represent categorical, discrete and continuous data.
  Choose and explain which representation to use in a given situation:
    - pie charts
    - line graphs and time series graphs
    - scatter graphs
    - infographics.
- Interpret data, identifying patterns, trends and relationships, within and between data sets, to answer statistical questions. Discuss conclusions, considering the sources of variation, including sampling, and check predictions.

Starting point

Do you remember…

- how to draw and interpret simple pie charts and line charts?

  For example, 60 people name their favourite item of salad.

Vegetable	Tomato	Lettuce	Cucumber	Pepper
Frequency	24	18	10	8

  Draw a pie chart to show the results.

- how to draw and interpret simple scatter graphs and infographics?

  For example, draw a line of best fit on this scatter diagram.

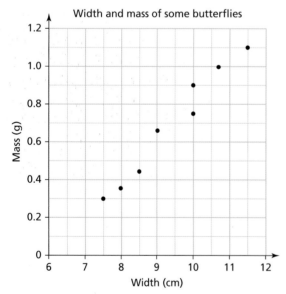

This will also be helpful when…

- you compare two distributions represented graphically
- you explore correlation.

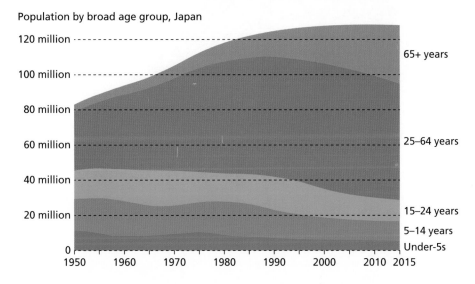

17.0 Getting started

Real data question The diagram shows changes in the population of Japan between 1950 and 2015.

Population by broad age group, Japan

Source: Hannah Ritchie (2020) - "Age Structure". Published online at OurWorldInData.org. Retrieved from: 'https://ourworldindata.org/age-structure' [Online Resource].

Discuss how the population of Japan has changed during this time.

17.1 Pie charts

Worked example 1

The table shows how 80 students travel to school.

Draw a pie chart to show the information.

Method of travel	Number of students
walk	38
car	22
bus	10
train	4
bicycle	6

Walk $\frac{38}{80} \times 360 = 171°$

Car $\frac{22}{80} \times 360 = 99°$

Bus $\frac{10}{80} \times 360 = 45°$

Train $\frac{4}{80} \times 360 = 18°$

Bicycle $\frac{6}{80} \times 360 = 27°$

Method of travel to school

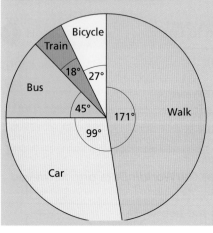

Find the fraction of the students who use each method of travel.

Work out the angle for each sector by multiplying this fraction by 360°.

Measure the angle of each sector using a protractor.

Remember to label each sector.

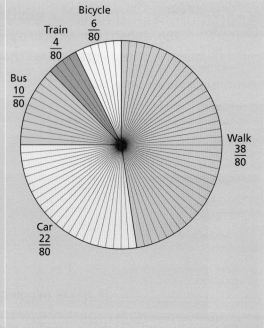

Worked example 2

Seamus measures the lengths of the tails of a random sample of 96 zebra. He categorises the length of each tail as very short, short, medium, long or very long. The pie chart shows his results.

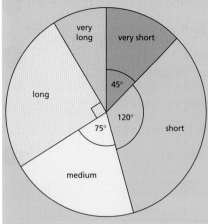

a) Find what fraction of the sample of zebra have a very long tail.

b) Calculate the number of zebras in the sample that have a medium length tail.

c) Seamus says that 60% of zebra in the population have a long tail. Is he likely to be correct? Give reasons for your answer.

a) $45° + 120° + 75° + 90° = 330°$

The angle for the 'very long' sector is $360° - 330° = 30°$

The fraction of zebra with 'very long' tails is $\frac{30}{360} = \frac{3}{36} = \frac{1}{12}$

The sum of the angles must be 360°. Use this fact to find the angle for the 'very long' sector.

Write this as a fraction of 360° and simplify.

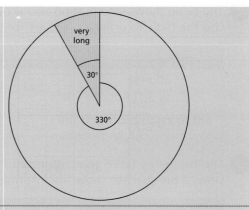

b) The fraction of the sample with 'medium' tails is

$\frac{75}{360} = \frac{5}{24}$

The number with a medium length tail is $\frac{5}{24}$ of 96

$= 5 \times 96 \div 24$

$= 20$

Find the fraction of zebra with 'medium' tails by writing the angle as a fraction of 360.

The total number of zebras in the sample is 96. So, find this fraction of 96.

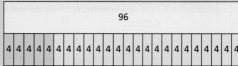

c) The fraction of zebras in the sample with 'long' tails is $\frac{90}{360} = \frac{1}{4}$ This is 25%

Seamus is likely to be incorrect. The percentage of zebra in the population with 'long' tails is likely to be about 25%.

First find the fraction of the sample that have 'long' tails. Convert this fraction to a percentage.

The percentage of the whole population with 'long' tails should be approximately the same as the percentage in the sample.

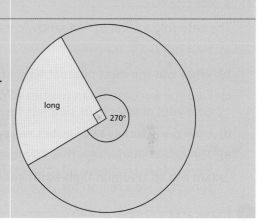

Exercise 1

1 The table shows information about 150 passengers travelling on a train.
Use the table to draw a pie chart for this data.

Passengers	Frequency
Men	45
Women	69
Boys	21
Girls	15

2 The table shows information about hair colour in a group of students.
Use the table to draw a pie chart for this data.

Hair colour	Percentage
Blonde	28
Brown	48
Black	24

3 The frequency table gives information about the favourite type of film for 440 people.
Draw a pie chart to show the information.

Type of film	Frequency
Action	56
Comedy	135
Romance	212
Science fiction	37

4 The pie chart shows the sales of different types of sandwich in a café one day.

The café sold a total of 144 sandwiches on that day.

a) Find the angle for Other.

b) What was the most popular type of sandwich?

c) Find the fraction of sandwiches sold that were cheese salad.
Give your answer in its simplest form.

d) How many turkey sandwiches were sold?

e) The café manager says that 40 more egg salad sandwiches were sold than cheese salad sandwiches. Is the manager correct? Give a reason for your answer.

Types of sandwiches sold

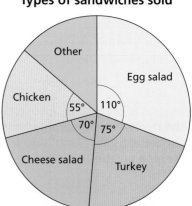

5 Katya asks a sample of 840 customers of a clothes shop what design of shirt they prefer. Her results are shown in the pie chart.

a) Write down the modal design.

b) Find the fraction of the sample whose preferred design of shirt was **not** plain. Give your answer in its simplest form.

c) Calculate the number of people in the sample whose preferred design of shirt was checked.

d) Calculate an estimate for the percentage of all the shop's customers whose preferred design of shirt is spotted.

Preferred design of shirt

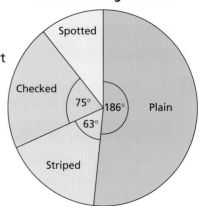

6 Real data question

a) The table shows the area of different types of land on Earth.

Type of land	Area (million km²)
Cultivated land	15.6
Forest	37.3
Grassland and woodland	45.9
Unproductive land	27.6
Settlement	1.5
Inland water	2.4

Source: Food and Agriculture Organization of the United Nations, 2011, The state of the world's land and water resources for food and agriculture (SOLAW) – Managing systems at risk. Reproduced with permission.

Draw a pie chart to show this information.

b) The total area of land occupied by middle-income countries is 68.6 million km².

Rita draws a pie chart to show the proportion of land in middle income countries that is forest land.

Find the area of Forest land in middle income countries.

Land use in middle-income countries

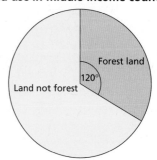

7 The pie chart shows the colours of the shirts sold in a shop one week.

The shop sold 64 black shirts.

How many more blue shirts than white shirts were sold?

Colour of shirts sold

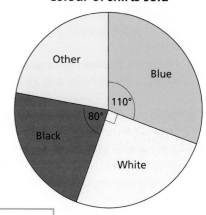

▼ **Thinking and working mathematically activity**

A shop has 600 pairs of jeans.

35% of the jeans are black.

The modal colour of the jeans is blue.

The shop has twice as many blue jeans as grey jeans.

All the remaining jeans in the shop are white.

- Draw a possible pie chart to show the colours of jeans in the shop.
- Write two more sentences to describe the colours of jeans in the shop.

Worked example 2

The map shows information about the percentage of adults who are literate (meaning they can read and write) in African countries in 2015.

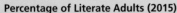

Percentage of Literate Adults (2015)

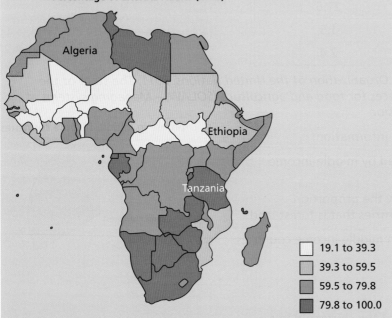

☐	19.1 to 39.3
▨	39.3 to 59.5
▩	59.5 to 79.8
▧	79.8 to 100.0

Source: © December 2019 by PopulationPyramid.net.

a) Maxine says that the percentage of adults who were literate in Tanzania in 2015 is 75%. Could she be correct? Explain your answer.

b) Compare the percentage of adults who were literate in Algeria and Ethiopia in 2015.

a) The percentage of adults who were literate in Tanzania is in the interval 79.8 to 100. So, Maxine cannot be correct.	Find Tanzania on the map and note its colour. Look at the key to find the interval that its percentage is within.
b) The percentage of adults who were literate in Algeria is greater than the percentage in Ethiopia.	Find both countries on the map. The percentage of adults who were literate in Algeria is in the interval 59.5 to 79.8% The percentage of adults who were literate in Ethiopia is in the interval 39.3 to 59.5% The percentage must be greater in Algeria.

> **Did you know?**
> This type of infographic is sometimes called a choropleth map.

> **Discuss**
> What other comparisons can you make between the literacy rates in different countries in Africa?

1 Real data question The infographic shows the number of rhinos poached in South Africa between 2006 and 2016.

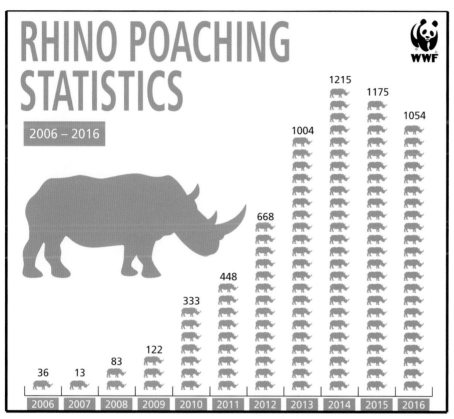

Source: WWF South Africa https://wwf.panda.org/wwf_news/?293410/
South%5FAfrica%5Frhino%5Fpoaching%5Ffigures%5F2016.

a) Describe the pattern in the number of rhinos being poached in South Africa during this time period.

b) Find the increase in the number of rhinos poached between 2010 and 2013.

c) How many times more rhinos were poached in 2014 compared with 2009?

d) There were about 20 000 rhinos in South Africa in 2016.
   Find the percentage of rhinos that were poached in that year.

2 Real data question The list shows some of the countries in South America and the percentage of the population in each country that is aged 65 years or older.

Argentina	11.2%	Guyana	6.7%
Bolivia	7.3%	Paraguay	6.6%
Brazil	9.3%	Peru	8.4%
Chile	11.9%	Suriname	7.0%
Colombia	8.8%	Uruguay	14.9%
Ecuador	7.4%	Venezuela	7.6%
French Guiana	5.3%		

Source: United Nations, Department of Economic and Social Affairs, Population Division (2019).
World Population Ageing 2019: Highlights (ST/ESA/SER.A/430).

a) Create a grouped frequency table to summarise the data.

b) Use your groupings to draw an infographic to show the data. Use a map of South America, like the one below, to help you.

3 The infographic shows sales in different departments in a store in 2020.

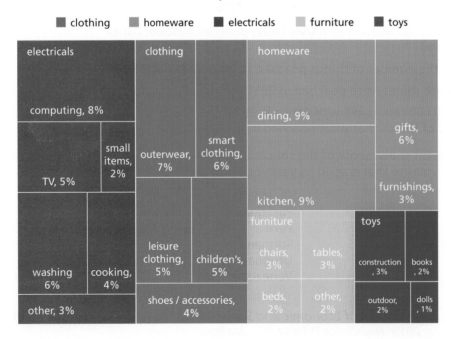

a) Write down the department which had the highest total sales.

b) Kieran says, 'The store's sales of leisure clothing were greater than its sales of smart clothing.' Is Kieran correct? Give a reason for your answer.

c) Calculate the percentage of sales for the furniture department.

d) The store's total sales were $90 million. Calculate the total sales for gifts.

4 **Real data question** The infographic shows the number of cases of measles in some European countries in 2018.

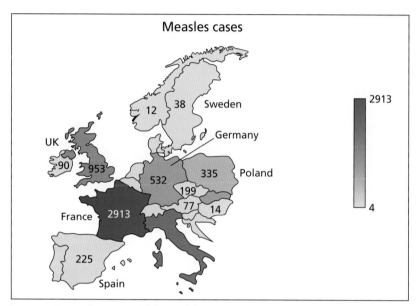

Source: Data from World Health Organization, 2019.

a) Name the country on the map with the highest number of measles cases.

b) Write down the number of cases of measles in Spain.

c) Compare the number of measles cases in the UK and Poland.

> **Think about**
>
> This diagram shows the number of cases of measles in each country. Another way to show the data would be to calculate the number of cases per million people living in the country. Why might this be a better way to compare the countries?

Thinking and working mathematically activity

Real data question The diagram shows the annual change in the area of forest in different continents of the world between 1990 and 2010.

Annual change in forest area by region, 1990-2010

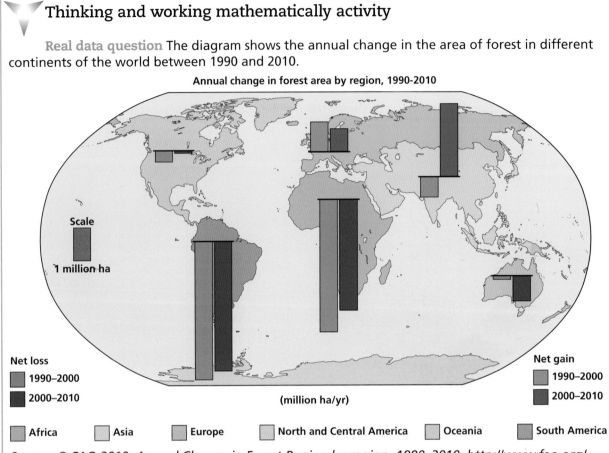

Net loss
- 1990–2000
- 2000–2010

Net gain
- 1990–2000
- 2000–2010

(million ha/yr)

Scale
1 million ha

☐ Africa   ☐ Asia   ☐ Europe   ☐ North and Central America   ☐ Oceania   ☐ South America

Source: © FAO 2010, Annual Change in Forest Region by region, 1990–2010, http://www.fao.org/forestry/fra/69406/en/ 16/09/2020.

- Discuss some conclusions that can be drawn from the graph.
- Discuss how successfully you think the graph is showing the information.
- How else could the information be displayed?

17.3 Trends and relationships

Key terms

A **time series graph** is a particular type of line graph where time is plotted along the horizontal axis. A time series graph can show **trends** over time – data may show an increasing trend or a decreasing trend.

Many time series graphs have years plotted on the horizontal axis. A year may be subdivided into quarters. The first quarter of the year is the first three months (January to March), the second quarter is the next three months and so on.

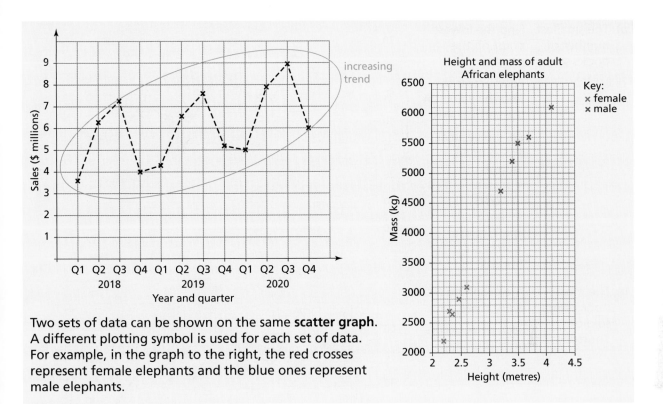

Two sets of data can be shown on the same **scatter graph**.
A different plotting symbol is used for each set of data.
For example, in the graph to the right, the red crosses
represent female elephants and the blue ones represent
male elephants.

Height and mass of adult African elephants

Key:
× female
× male

Discuss

How do the height and mass of male and female African elephants compare?

Worked example 3

A small bookshop opens three days each week, Monday, Wednesday and Thursday.

The graph shows the number of books sold by the bookshop over four weeks.

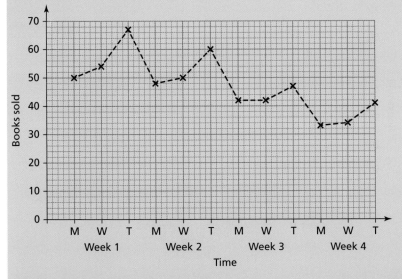

a) Find the smallest number of books sold on any day during this time period.

b) On which day of every week were the most books sold?

c) Describe the trend in the number of books sold over these four weeks.

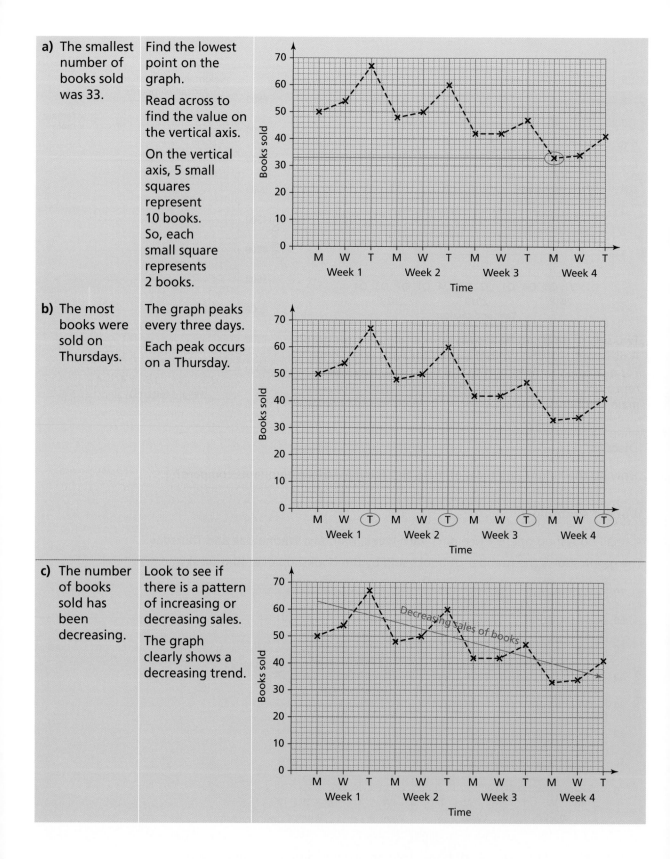

a) The smallest number of books sold was 33.

Find the lowest point on the graph.

Read across to find the value on the vertical axis.

On the vertical axis, 5 small squares represent 10 books. So, each small square represents 2 books.

b) The most books were sold on Thursdays.

The graph peaks every three days.

Each peak occurs on a Thursday.

c) The number of books sold has been decreasing.

Look to see if there is a pattern of increasing or decreasing sales.

The graph clearly shows a decreasing trend.

Worked example 4

Ten men took part in two quizzes, a sports quiz and a music quiz.

The scatter graph shows the number of points scored by each person in the two quizzes.

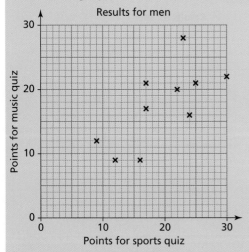

a) Find the median number of points scored by men in the sports quiz.

b) Describe the relationship between the points scored by these men in the two quizzes.

c) Adam scored 20 points in the sports quiz but did not take the music quiz. By drawing a line of best fit on the scatter graph, predict what his score might have been in the music quiz. How accurate do you think your prediction is likely to be?

Ten women also took the two quizzes. Their results are shown in this scatter graph.

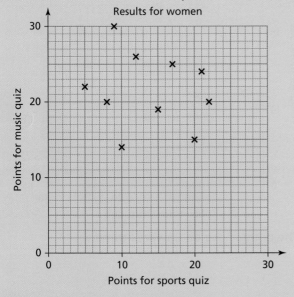

d) Compare the points scored by men and women on the music quiz.

> **Think about**
>
> If you want to compare two scatter graphs, it is sensible to draw them using the same scales. Why is this sensible?

a) The median is half-way between the 5th and the 6th data values.

The 5th value is 17 points.

The 6th value is 22 points.

So, the median is 19.5 points.

Remember that when there are *n* data values, the median value is in position $\frac{n+1}{2}$

Look at the horizontal coordinates of each of the points plotted.

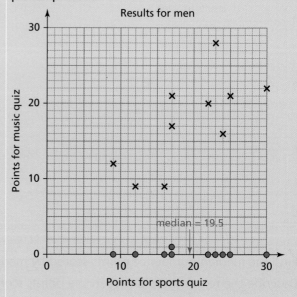

b) Men who did well in the sports quiz also tended to do well in the music quiz.

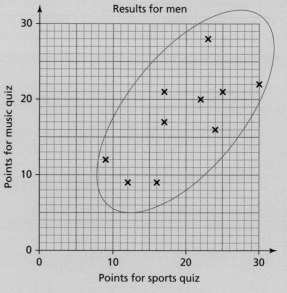

The points have an upwards trend.

c)

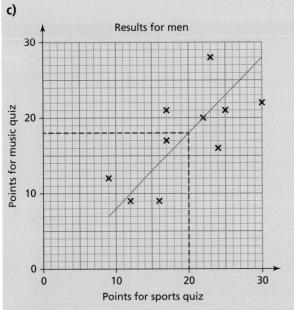

Results for men

An estimate for Adam's score in the music quiz is 18 points.

The estimate is not likely to be very reliable as the points show a fair degree of scatter about the line of best fit.

When positioning the line of best fit, try to ensure that:

- it follows the trend in the data
- the points are roughly balanced either side of the line.

To predict Adam's score in the music quiz:

- find 20 on the 'sports quiz' axis
- draw a line vertically up to the line of best fit and then horizontally across to the 'music quiz' axis.

d) Overall, women seemed to score higher than men on the music quiz.

Compare the values that the men and women scored on the music quiz by looking at the vertical plots of the points.

The points for women tend to be slightly higher up the graph than the points for men.

More women than men, for example, scored at least 20 points on the music quiz.

Exercise 3

The diagram shows the population of Nigeria from 2012 to 2018.

a) Write down the population of Nigeria in 2012.

b) Write down the first year in which the population was greater than 190 million.

c) Describe the trend in the population of Nigeria.

d) Find the difference in the population between 2015 and 2018.

e) Predict the population in 2019. Show how you found your answer.

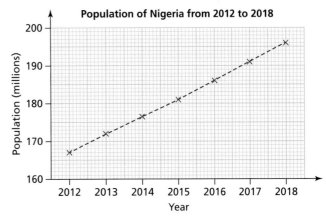

Population of Nigeria from 2012 to 2018

2 The table shows Nikolas' electricity bill each quarter between 2018 and 2020.

Year	2018				2019				2020			
Quarter	Q1	Q2	Q3	Q4	Q1	Q2	Q3	Q4	Q1	Q2	Q3	Q4
Bill ($)	250	212	197	237	265	220	205	244	278	254	218	261

a) Draw a time series graph to show Nikolas' electricity bills.

b) In which quarter is Nikolas' electricity bill the lowest each year?

c) Describe the trend in Nikolas' electricity bills.

3 The line graph shows the number of students studying History and Geography at a school.

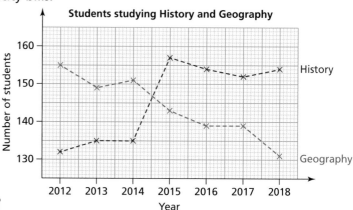

a) Describe the trend in the number of students studying Geography at the school.

b) How many more students studied History in 2018 than in 2012?

c) Between which two years did the number of students studying History rise by the greatest amount?

d) Find how many more students studied Geography than History in 2013.

4 The line graph shows the percentage of the population of Australia, Japan and Germany living in towns between 1960 and 2010.

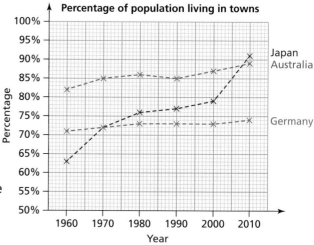

a) Which country had the highest percentage of people living in towns in 2000?

b) Estimate the percentage of the population living in towns in Germany in 1980.

c) In which year did Germany and Japan have approximately the same percentage?

d) Find the difference in the percentages between Australia and Japan in 1990.

e) Which country saw the greatest change in the population living in towns between 1960 and 2010?

f) Estimate the percentage of the population living in towns in Australia in 2005. Explain why your answer is an estimate.

5 A small cinema opens on Sunday, Monday and Tuesday each week.
It shows one film on each of these days. The time series graph shows
the number of people watching the film over a four-week period.

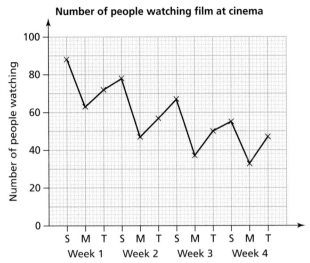

Number of people watching film at cinema

a) How many people watched the film on Monday in Week 2?

b) Describe any patterns you see in the data.

c) Find how many fewer people went to the cinema on Sunday in Week 4
compared with Sunday in Week 1.

d) Predict the number of people watching the film at the cinema on Sunday in Week 5.
Explain how you worked out your answer.

6 Jeremy has been asked to investigate this question:

*How has the number of children at a music school learning to play different musical instruments
changed since 2010?*

He collects this information shown in the graph.

Write some conclusions related to the question that Jeremy was investigating.

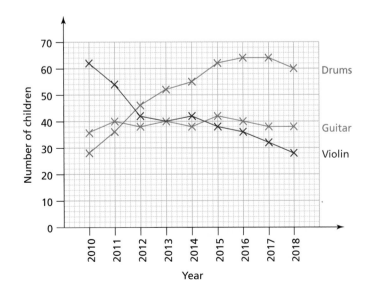

Thinking and working mathematically activity

Real data question The tables show the amount of plastics that were produced and the percentage of plastic that was recycled in different years.

Global plastic production

Year	Production (million tonnes)
1955	4
1965	17
1975	46
1985	90
1995	156
2005	263
2015	381

Recycling of plastic

Year	Percentage of plastic recycled
1985	0
1990	2
1995	5.5
2000	9
2005	12.5
2010	16
2015	19.5

Source: Hannah Ritchie (2018) - "FAQs on Plastics". Published online at OurWorldInData.org. Retrieved from: 'https://ourworldindata.org/faq-on-plastics' [Online Resource].

- Produce graphs showing the information in the tables.
- Produce a poster to show changes in global plastic production and recycling rates.
- Write some descriptions of your graphs. What trends do you notice?

7 The table shows the length and mass of some letters Aabhan receives through the post.

Length (cm)	32	11	22	31	25	28	16	20	18
Mass (grams)	105	24	44	86	63	42	45	64	32

a) Draw a scatter graph to show this information.

b) Describe the relationship between the length and mass of the letters.

c) Draw a line of best fit on your scatter graph.

d) A different letter has a length of 23 cm. Use your line of best fit to predict the mass of the letter.

8 A Mathematics teacher investigates whether students who are absent more from school score less well in the end of year exam.

The table shows her data.

Percentage absence	9	6	0	18	5	4	1	2	13	8	12	14
Exam score (%)	73	60	77	32	56	52	89	64	43	47	37	54

a) Draw a scatter graph to show the data.

b) Make a conclusion about how absence from school is related to exam score.

c) Gabriella scored 50% in the end of year exam. Draw a line of best fit on your scatter graph and predict Gabriella's percentage absence.

9 Sheree buys some cookery books and some sports books. She records the cost and the number of pages in each book. The scatter diagram shows her results.

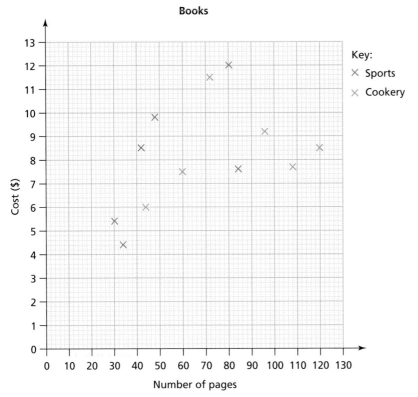

Key:
× Sports
× Cookery

a) Write down how many sports books Sheree buys.

b) Write down the cost of the most expensive cookery book Sheree buys.

c) Find the number of cookery books bought with more than 80 pages.

d) Calculate the range in the costs of the sports books Sheree buys.

e) Compare the number of pages in Sheree's sports books with the number of pages in her cookery books.

10 A hotel keeps records showing how many nights its customers stay and the total cost of each stay.

The tables show this information for seven customers in June and seven customers in December.

June

Number of nights	3	5	7	6	8	6	4
Cost ($)	240	310	480	460	555	440	350

December

Number of nights	4	2	1	3	4	2	3
Cost ($)	300	150	65	145	245	130	190

a) Plot this information on a single graph. Use different types of plotting symbols to represent the data in each table. Remember to include a key.

b) Explain why your graph is an appropriate way to show the data.

c) Describe the relationship between the number of nights and the cost.

d) Compare the data for June and December.

11 The tables show the number of scratches on some cars and vans.
The number of miles (to the nearest thousand) each car has driven is also shown.

Cars

Number of miles (thousands)	11	17	28	34	42	58	76
Number of scratches	2	8	11	7	9	17	21

Vans

Number of miles (thousands)	8	23	34	45	57	68	79
Number of scratches	6	17	15	21	29	27	35

a) Draw a scatter graph showing the information for cars. Add a line of best fit.

b) Draw a separate scatter graph showing the information for vans. Add a line of best fit.

> **Tip**
>
> In part **b)**, you should use the same scales as you used in part **a)**.

c) Use the lines of best fit to estimate the number of scratches on a car and the number of scratches on a van which has driven 50 000 miles. Compare these values.

Consolidation exercise

1 Technology question The table shows information about land use in a country.

Type of land	Area (thousands km²)
Agricultural	312
Forest	119
Other	136

Use a spreadsheet to draw a pie chart to show the information.

2 The table shows the number of visitors to an island at different times of a year.

Year	2018			2019			2020		
Months	Jan – Apr	May – Aug	Sep – Dec	Jan – Apr	May – Aug	Sep – Dec	Jan – Apr	May – Aug	Sep – Dec
Visitors (thousands)	140	410	150	150	435	165	180	460	185

a) Draw a time series graph to show the information.

b) Comment on any patterns shown in the data.

3 Susie records the number of cups of coffee and the number of pieces of cake she sells on eight different days.

Number of cups of coffee sold	108	96	97	135	132	108	124	101
Number of pieces of cake sold	43	34	27	46	51	35	39	24

a) Draw a scatter graph to show the information.

b) Describe the relationship between the number of cups of coffee sold and the number of pieces of cake sold.

c) On a different day, Susie sells 130 cups of coffee. Predict how many pieces of cake she sells on this day. Show clearly how you worked out your answer.

4 Charlie records the temperature outside at midday and midnight on 12 different days He draws this diagram to show his results.

What has Charlie done wrong in drawing his scatter graph?

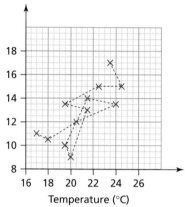

5 **Real data question** These infographics show the population of countries and continents in 2020 and the forecast populations for 2100.

List of countries ordered by their population size, 2020
Total: **7,794,798,729**

Source: © December 2019 by PopulationPyramid.net.

List of countries ordered by their population size, 2100
Forecast: **10,875,393,719**

Source: © December 2019 by PopulationPyramid.net.

a) Write down the country forecast to have the greatest population in 2100.

b) How is the population of Africa forecast to change between 2020 and 2100?

c) Make some other comparisons from the charts.

End of chapter reflection

You should know that...	You should be able to...	Such as...
In a pie chart, each category of data is represented by a sector of a circle.	Draw and interpret a pie chart when the total frequency is not a factor of 360°.	Draw a pie chart to show the type of magazine that 50 people most like to read.

Type of magazine	Number of people
News	8
Fashion	14
Sport	16
Technology	12

Infographics are a visual way to represent a set of data. They are used to make a set of data easier to interpret.	Draw and interpret a range of different infographics.	A country has six forests. The infographic shows the area that each forest covers. 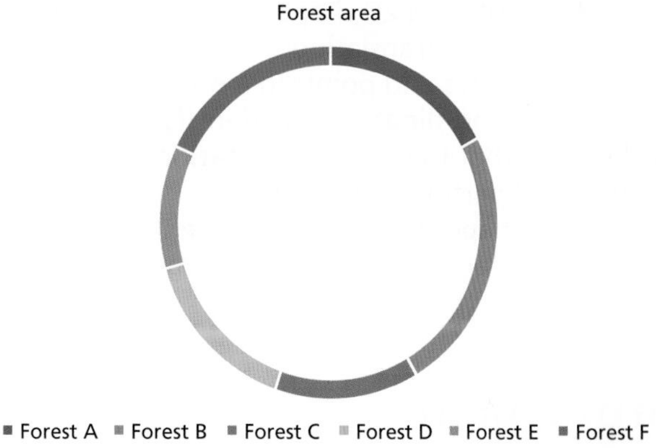 Forest area ■ Forest A ■ Forest B ■ Forest C ■ Forest D ■ Forest E ■ Forest F Write down the forest that covers the most area.
Line graphs and time series graphs can show trends in data over time.	Draw a line graph or time series graph and comment on the trend.	Draw a line graph to show the number of members of a tennis club. | Year	Number of members
---	---
2013	47
2014	61
2015	69
2016	73
2017	74	 Comment on the trend in the data.
A scatter graph shows relationships between variables.	Draw and interpret a scatter graph.	The table shows the size and cost of some packs of pasta. | Size of pack (g)	Cost ($)
---	---
500	0.60
1000	1.80
300	1.25
1000	1.10
400	1.05
250	0.55
750	1.35	 Draw a scatter graph to show the information. Draw a line of best fit. Use your line of best fit to predict the cost of a 600 g pack of pasta.

Transformations

You will learn how to:

- Translate points and 2D shapes using vectors, recognising that the image is congruent to the object after a translation.
- Reflect 2D shapes and points in a given mirror line on, or parallel to, the *x*- or *y*-axis, or *y* = ± *x* on coordinate grids. Identify a reflection and its mirror line.
- Understand that the centre of rotation, direction of rotation and angle are needed to identify and perform rotations.
- Enlarge 2D shapes from a centre of enlargement (outside or on the shape) with a positive integer scale factor. Identify an enlargement and scale factor.

Starting point

Do you remember…

- how to find the image of a point or a shape under a translation?

  For example, find the image of the point (–6, 7) when it is translated 5 units to the right and 2 squares down.

- how to reflect a shape in one of the axes on a coordinate grid?

  For example, shape D is a reflection of shape C in the *y*-axis.

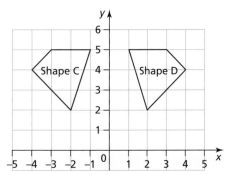

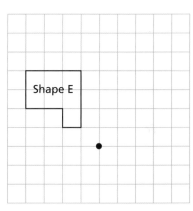

- how to rotate a shape around a point?

  For example rotate shape E by 90° clockwise around the point.

- how to enlarge a shape?

  For example, enlarge shape E by scale factor 2

- that two shapes are congruent when they are exactly the same shape?

  For example, the image of a reflection is congruent to the object.

- how to recognise the equations of horizontal and vertical lines?

  For example, the line *x* = 2 is a vertical line parallel to the *y*-axis and the line *y* = –1 is a horizontal line parallel to the *x*-axis.

18.0 Getting started

Sophia wants to cover her patio using rectangular brown and white tiles.

Here is one design she considers.

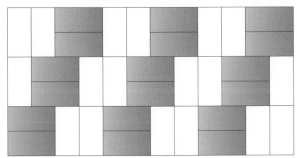

Create your own designs to cover the patio by using the following suggestions:

- Creating a design that can be reflected, horizontally as well as vertically.
- Creating a design that can be rotated 90° or 180° so that the whole patio is covered.
- Creating a design using one colour tile and a white tile placed by the same translation.

18.1 Translations

Key terms

A **transformation** changes the position and/or the size of a shape.

Translations can be described using **vectors**.
A vector is represented by the combination
of a horizontal movement and a vertical
movement.

For example,

Vector $\begin{pmatrix} 3 \\ 1 \end{pmatrix}$ shows how triangle A has moved to triangle B:

    3 units horizontally right
    1 unit vertically up

Vector $\begin{pmatrix} 1 \\ -5 \end{pmatrix}$ shows how triangle A has moved to triangle C:

    1 unit horizontally right
    5 units vertically down

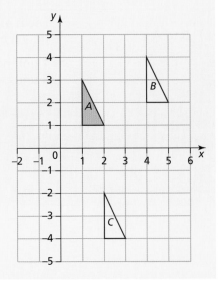

Worked example 1

Use vectors to describe the translations of the following triangles.

a) A to C

b) B to A

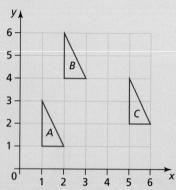

a) A to C is $\begin{pmatrix} 4 \\ 1 \end{pmatrix}$	Focus on one vertex on A and see how it moves to C: 4 units horizontally right 1 unit vertically up	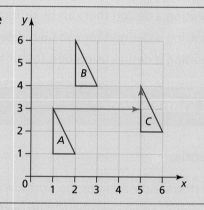
b) B to A is $\begin{pmatrix} -1 \\ -3 \end{pmatrix}$	Focus on one vertex on B and see how it moves to A: 1 unit horizontally left 3 units vertically down	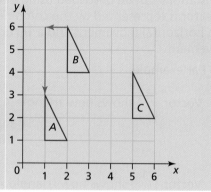

1 Using the diagram below, describe these translations with vectors.

a) A to B   b) A to C   c) A to D   d) C to B   e) C to E

f) E to B   g) B to A   h) C to D   i) E to A   j) D to B

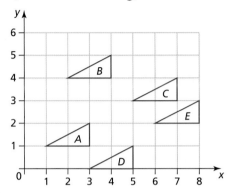

2 a) Copy this triangle onto squared paper.

b) Draw the image of A after a translation with vector $\begin{pmatrix} 2 \\ 3 \end{pmatrix}$. Label this P.

c) Draw the image of A after a translation with vector $\begin{pmatrix} -1 \\ 2 \end{pmatrix}$. Label this Q.

d) Draw the image of A after a translation with vector $\begin{pmatrix} 3 \\ -2 \end{pmatrix}$. Label this R.

e) Draw the image of A after a translation with vector $\begin{pmatrix} -2 \\ -4 \end{pmatrix}$. Label this S.

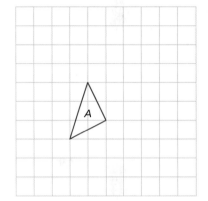

3 Using your diagram from question 2, describe the translation that will move:

a) P to Q   b) Q to R   c) R to S   d) S to P

e) R to P   f) S to Q   g) R to Q   h) P to S

4 Copy and complete this table.

Description of translation	Vector
2 units right and 5 units up	
3 units left and 2 units down	
	$\begin{pmatrix} 3 \\ -4 \end{pmatrix}$
	$\begin{pmatrix} -2 \\ 6 \end{pmatrix}$
7 units left and 5 units up	

5 Write down the image of the point $P(3, 2)$ after a translation of:

a) $\begin{pmatrix} 4 \\ 3 \end{pmatrix}$
b) $\begin{pmatrix} 5 \\ -1 \end{pmatrix}$
c) $\begin{pmatrix} -2 \\ 6 \end{pmatrix}$
d) $\begin{pmatrix} -4 \\ -2 \end{pmatrix}$

6 Write down the image of the point $(-2, -4)$ after a translation of:

a) $\begin{pmatrix} 2 \\ -3 \end{pmatrix}$
b) $\begin{pmatrix} -3 \\ 5 \end{pmatrix}$
c) $\begin{pmatrix} -5 \\ -7 \end{pmatrix}$
d) $\begin{pmatrix} 6 \\ 2 \end{pmatrix}$

7 Dominique wrote a vector of a translation as (3, 6).

Explain what is wrong about Dominique's answer.

8 Write down the vector that describes a translation moving the point $(-1, 1)$ to the points:

a) (3, 6)
b) (5, -3)
c) (-4, -5)

9 The image of the point $(2, -5)$ under a translation is $(5, 2)$.

Find the coordinates of the image of (1, 1) under the same translation.

10 State whether each statement is true or false.

a) After a translation, an image is congruent to the object.
b) Under a translation, each point moves with the same vector.
c) A rectangle can be translated so that it becomes a square.
d) A vector can look the same as a coordinate.

11 Match each pair of points to the matching translation.

| (3, 4) moves to (3, 8) |
| (-3, 2) moves to (0, 4) |
| (4, -2) moves to (2, 5) |
| (1, -4) moves to (-2, -6) |

$\begin{pmatrix} 3 \\ 2 \end{pmatrix}$

$\begin{pmatrix} -2 \\ 7 \end{pmatrix}$

$\begin{pmatrix} -3 \\ -2 \end{pmatrix}$

$\begin{pmatrix} 0 \\ 4 \end{pmatrix}$

12 $ABCD$ is a parallelogram with coordinates $A(2, 3)$, $B(7, 3)$ and $C(4, 5)$.

It is translated so that point A moves to $(0, -1)$

a) Write down the translation vector.

b) Find the coordinates of where point D has been translated to.
Show how you worked out your answer.

Thinking and working mathematically activity

Look at the triangle ABC and its image $A'B'C'$

The triangle ABC has been translated to give triangle $A'B'C'$.

- What is the vector that translates A to A'?

- What is the vector that translates A' to A?

- What is the connection between these vectors?

- Is the same connection true for the vectors translating B to B' and B' to B?

- Is this connection true for all translations and the reverse translation?

- Can you write a generalisation or a rule from this investigation?

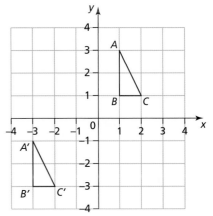

18.2 Reflections

Worked example 2

a) Reflect shape A in the line $y = 1$.
   Label the image B.

b) Reflect shape A in the line $y = x$.
   Label the image C.

c) Identify the line that reflects shape P to shape Q.

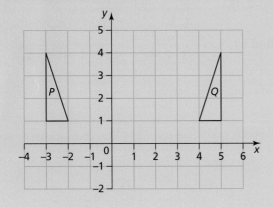

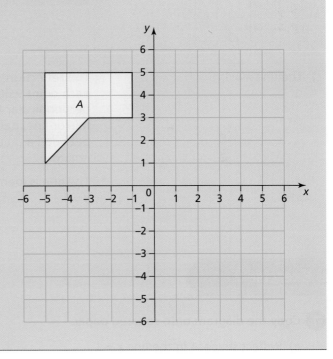

a)

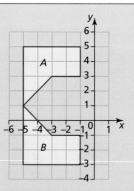

Draw the mirror line $y = 1$ then reflect each of the corners of shape A in the mirror line and join them up.

Use lines that are perpendicular to the mirror line when reflecting.

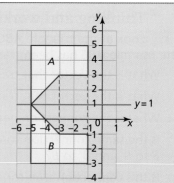

b) The shapes B and C are congruent to A.

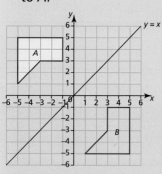

Draw the mirror line $y = x$. Reflect each corner of A in this mirror line.

Count diagonally the number of squares from each corner to the mirror line. Repeat this on the opposite side.

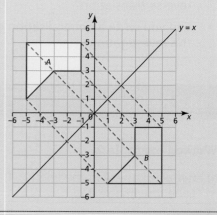

c) The line $x = 1$

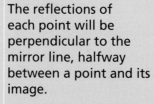

The reflections of each point will be perpendicular to the mirror line, halfway between a point and its image.

The line is identified as $x = 1$

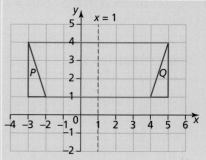

Exercise 2

1 Copy the diagram onto squared paper.

a) Reflect shape A in the line $x = 1$
Label the image B.

b) Reflect shape A in the line $y = -1$
Label the image C.

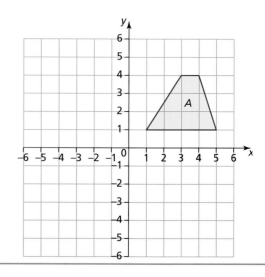

2 Copy the diagram onto squared paper.

a) Reflect shape *A* in the line *y* = 1
Label the image *B*.

b) Reflect shape *A* in the line *x* = 1
Label the image *C*.

c) Reflect shape *B* in the line *x* = −1
Label the image *D*.

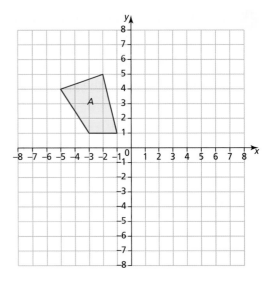

3 The diagram shows seven congruent triangles.

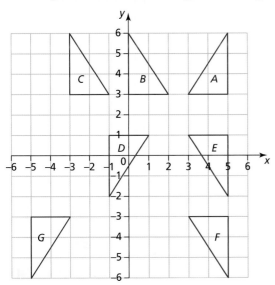

State the mirror line for the following reflections:

a) *G* to *F*   **b)** *A* to *C*   **c)** *A* to *E*   **d)** *A* to *F*   **e)** *E* to *D*   **f)** *A* to *B*

4 Eve looks at the diagram above and says, 'It looks like all reflections of a shape are translations of each other.' Is Eve correct? Explain your answer.

5 **a)** Draw a pair of axes, *x*-axis from −6 to 6, *y*-axis from −6 to 6.

b) Draw a triangle with coordinates (1, 2), (4, 3) and (3, 5). Label the triangle *A*.

c) Reflect triangle *A* in the line *y* = 1. Label the image *P*.

d) Reflect triangle *P* in the line *x* = −1. Label the image *Q*.

e) Reflect triangle *Q* in the line *y* = 1. Label the image *R*.

6 Copy the diagram onto squared paper.

a) Reflect shape A in the line $y = x$
   Label the image B.

b) Reflect shape A in the line $y = -x$
   Label the image C.

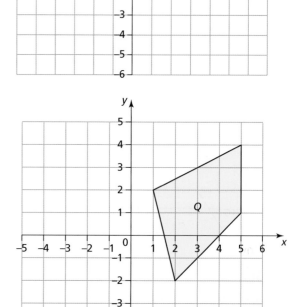

7 Copy this diagram onto squared paper.
   Reflect shape Q in the line $x = 2$

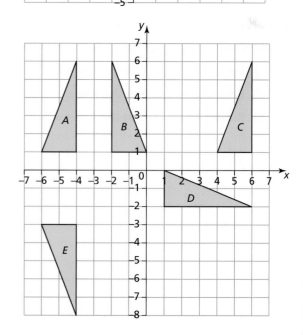

8 The diagram shows five triangles.

Write the equation of the mirror line for:

a) a reflection of shape A to shape B

b) a reflection of shape B to shape C

c) a reflection of shape A to shape E

d) a reflection of shape B to shape D

9 The diagram shows shape *ABCDEF*.

a) *ABCDEF* is reflected in a mirror line.
The images *A'* and *B'* of vertices *A* and *B* respectively are shown.

Copy the diagram and complete the image of the shape *ABCDEF* under the reflection.

b) *ABCDEF* is reflected in a different mirror line. The images *E"* and *D"* of vertices *E* and *D* respectively are shown.

Copy the diagram and complete the image of the shape *ABCDEF* under this reflection.

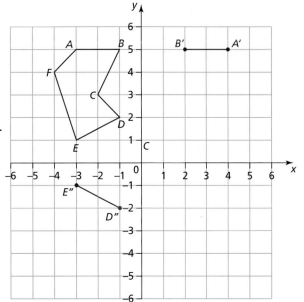

Thinking and working mathematically activity

- Draw a pair of axes, *x*-axis from –6 to 6, *y*-axis from –6 to 6
- Draw a triangle with coordinates *A*(–1, –4), *B*(5, –1) and *C*(5, 3)
- Draw the reflection of *ABC* in the line $y = x$.
- Investigate the connection between the coordinates of *A*, *B* and *C* and the coordinates of their images.
- Will the same connection hold if any point is reflected in this same mirror line? Investigate.
- Now consider what happens when you reflect a shape in the line $y = -x$. Investigate the relationship between the coordinates of an object and an image in this case.

18.3 Rotations

Key terms

A **rotation** is the turning of a shape.

You should state the angle and direction of the rotation as well as the centre point around which the shape turns.

For example, the triangle *ABC* has been rotated:

- 90° (the angle)
- clockwise (the direction)
- about the origin (0, 0) (the centre of rotation) to the triangle *A'B'C'*.

Each point has been rotated exactly the same way.

The object is congruent to its image.

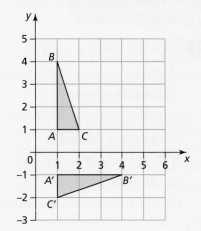

Worked example 3

a) Rotate shape *A*, 90° anticlockwise around the origin and label the image *B*.

b) Rotate shape *A*, 180° anticlockwise around the point (0, 1) and label the image *C*.

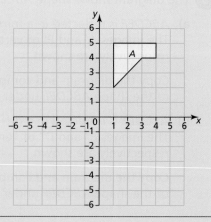

a) 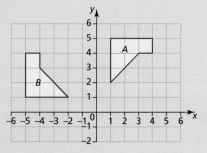	Trace shape A onto tracing paper. Put the point of your pencil on the centre of rotation, which here is the origin. Turn the tracing paper by a quarter turn anticlockwise.	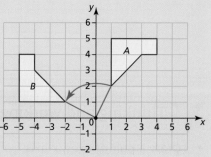
b) 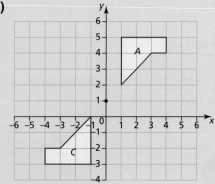	Trace shape A onto tracing paper. Put the point of the pencil on the centre of rotation, which here is the point (0, 1). Turn the tracing paper through half a turn.	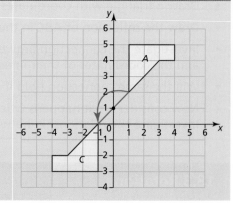

Exercise 3

1 Copy the diagram onto squared paper.

   a) Rotate shape *A*, 90° clockwise about (0, 0)

      Label the image *B*.

   b) Rotate shape *A*, 90° anticlockwise about (0, 0)

      Label the image *C*.

   c) Rotate shape *A*, 180° about (0, 0)

      Label the image *D*.

   d) Which shapes are congruent to *A*?

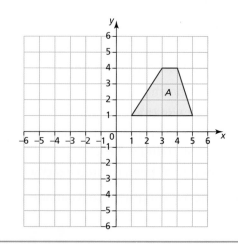

2 Copy the diagram onto squared paper.

   a) Rotate shape *A*, 90° clockwise about (1, 0)
      Label the image *B*.

   b) Rotate shape *A*, 90° anticlockwise about (0, 1)
      Label the image *C*.

   c) Rotate shape *A*, 180° about (0, 0)
      Label the image *D*.

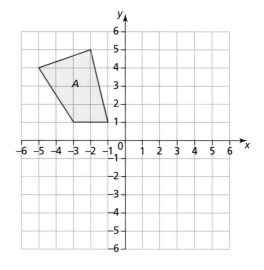

3 Copy the diagram and rotate the triangle by:

   a) 90° clockwise about (0, 0)

   b) 180° about (3, 3)

   c) 90° anticlockwise about (0, 2)

   d) 180° clockwise about (–1, 0)

   e) 90° clockwise about (–1, –1)

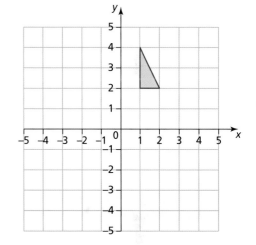

4 Complete the descriptions of the following rotations:

   a) Shape *A* to shape *B*.
      rotation of .... about (0, 1)

   b) Shape *D* to shape *C*.
      rotation of 90° anticlockwise about (..... , ...)

   c) Shape *E* to shape *C*.
      rotation of .... about (–1, –2)

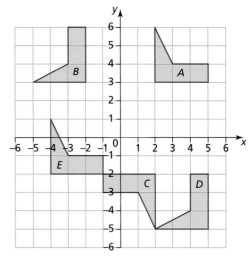

5 Which of these rotations are equivalent? Give a reason for your answer.

   270° clockwise   90° clockwise   90° anticlockwise   270° anticlockwise   180° clockwise

Thinking and working mathematically activity

The diagram shows a right-angled triangle R. The image of one side of R under a single transformation is shown.

- Copy the diagram and show all the possible positions of the completed image.
  There are three different possible positions.

- Decide whether each transformation is a translation, a reflection or a rotation.

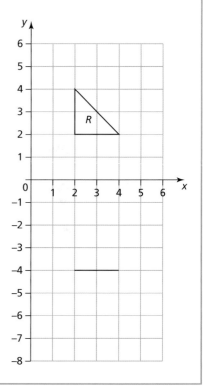

18.4 Enlargements

Key terms

An **enlargement** changes the size of a shape to give a similar image.

An enlargement has a **centre of enlargement** and a **scale factor.**

Every length of the enlarged shape can be calculated by: Original length × scale factor

The distance of each image point on the enlargement from the centre of enlargement will be:

   Distance of original point from centre of enlargement × scale factor

Worked example 4

The diagram shows a shape *L* drawn on a grid.

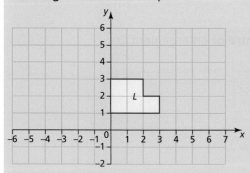

Find the image of *L* under each of these enlargements:

a) Scale factor 2 and centre of enlargement (6, 2)

b) Scale factor 3 and centre of enlargement (1, 2)

a)

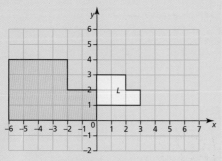

The scale factor for the enlargement is 2, so each vertex will be twice as far from the centre of enlargement as it is currently.

To get from the centre of enlargement to the corner of *L* marked with a red dot, move four left and one up. To get to the image, start from the centre of enlargement and move eight left and two up.

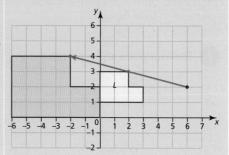

b)

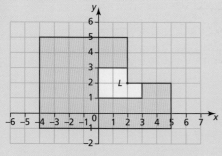

The scale factor for the enlargement is 3, so each vertex will be three times as far from the centre of enlargement as it is currently.

Draw a line from the centre of enlargement to each corner of *L* and extend to three times the length.

Ensure each line in the enlarged shape is three times as long as the original.

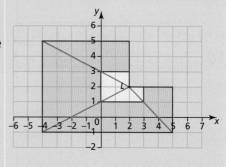

1 Copy each shape onto squared paper.

Enlarge each shape by the given scale factor from the centre of enlargement O.

a)
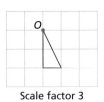

Scale factor 3

b)

Scale factor 2

c)
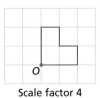

Scale factor 4

d)
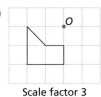

Scale factor 3

2 The diagram shows triangle T.

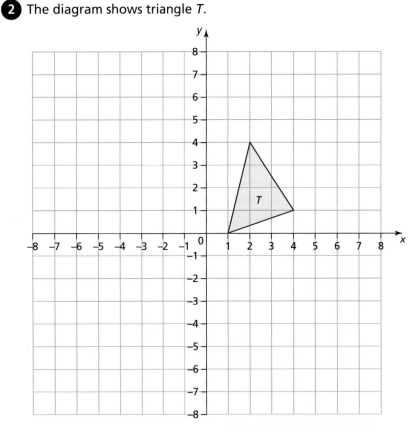

a) Enlarge T by scale factor 2, centre of enlargement (0, 1)

b) Enlarge T by scale factor 3, centre of enlargement (5, 3)

c) Enlarge T by scale factor 2, centre of enlargement (7, 0)

3 Four students look at this diagram.

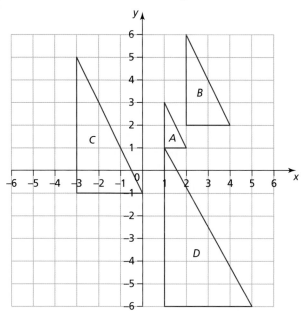

Are each student's comments below true or false? If false, explain your answer.

a) Mae says that triangle *B* is an enlargement of *A* with a scale factor of 2.

b) Kofi says that triangle *C* is an enlargement of *A* with a scale factor of 3.

c) Nia says that triangle *D* is an enlargement of *A* with a scale factor 4.

d) Theo says that (3, 2) is the centre of enlargement from *A* to *C*.

4 **Vocabulary question** Copy and complete the statements below using words from this box.

| scale factor | centre | angles | lengths | similar |

In an enlargement, the _____ of the original sides are all multiplied by the _____ to give the _____ of the sides in the image.

In an enlargement, all the _____ of the original shape are the same as those in the image.

After an enlargement, the two shapes are _____ .

The distance of each point on an enlargement from the _____ of enlargement is the distance of the original point from centre of enlargement multiplied by the _____ .

5 Explain how you could change one vertex of shape *N* in order to make *N* an enlargement of shape *M*.

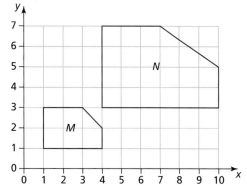

6 The diagram shows shape *L* and two enlargements.

a) What is the scale factor from shape *L* to shape *M*?

b) What is the scale factor from shape *L* to shape *N*?

c) Explain why shapes *L*, *M* and *N* are all similar.

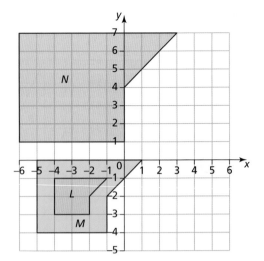

Thinking and working mathematically activity

Draw a pair of axes, $0 \leq x \leq 10$ and $0 \leq y \leq 10$

Draw a square *ABCD* with *A*(1, 1), *B*(2, 1), *C*(2, 0) and *D*(1, 0)

- Enlarge *ABCD* using a scale factor of 2, with the centre of enlargement at (0, 0)
  Write down the coordinates of the images of *A* and *B*.

- Repeat the above with a scale factor of 3

- Repeat the above with a scale factor of 4

- Repeat the above with a scale factor of 5

- Look at the pattern of the coordinates for the four images, and use them to
  predict the coordinates of the images of *A* and *B* for:

  - an enlargement using a scale factor of 6, with the centre of enlargement at (0, 0)

  - an enlargement using a scale factor of 10, with the centre of enlargement at (0, 0)

  - an enlargement using a scale factor of *n*, with the centre of enlargement at (0, 0)

- Investigate the different coordinate patterns from different centres of enlargement.

Consolidation exercise

1 Copy the diagram and then find the image of
T under each transformation.

a) Reflect *T* in the line $y = -1$. Label the image *U*.

b) Rotate *T* by 180° about the point (0, 0).
Label the image *V*.

c) Translate *T* with the vector $\begin{pmatrix} -3 \\ -1 \end{pmatrix}$.
Label the image *W*.

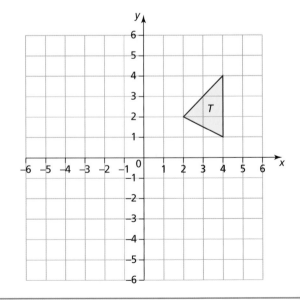

2 Copy the diagram and then find the image of
   P under each transformation.

   a) Reflect *P* in the line *x* = –2. Label the image *Q*.

   b) Rotate by 90° clockwise about the point (–2, 0).
   Label the image *R*.

   c) Enlarge *P* by scale factor 2, centre of enlargement
   (–1, 0). Label the image *S*.

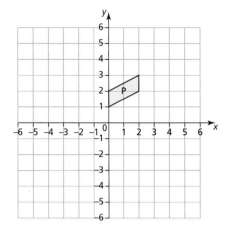

3 The diagram shows 9 triangles labelled *A* to *I*.

   Write down the letter of the image of *A* under
   these transformations.

   a) Reflection of *A* in the *x*-axis

   b) Translation of *A* by 4 units to the right

   c) Rotation of *A* by 180° centre (0, 0)

   d) Reflection of *A* in the line *x* = –2

   e) Rotation of *A* by 90° clockwise, centre (–3, –1)

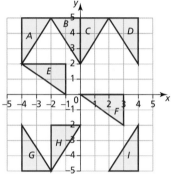

4 The diagram shows triangle *JKL*.

   a) The image of triangle *JKL* under a
   translation is *J'K'L'*. The position of *J'* is
   shown. Copy the diagram and
   complete the triangle *J'K'L'*.

   b) Sophia said the translation taking
   JKL to *J'K'L'* is (6, 7). State the error
   Sophia has made.

   c) The image of triangle *JKL* under a different
   transformation is triangle *J"K"L"*. The side
   J"L" is shown. Complete triangle *J"K"L"*.

   d) Lottie said the centre for this
   transformation is (5, –5). State the error
   Lottie has made.

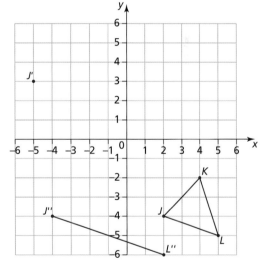

5 The diagram shows a right-angled triangle and the line *y* = *x*.

   a) Eve and Ollie try to find the image of the vertex with
   coordinates (2, 3) when it is reflected in the line *y* = *x*.
   Eve says the image is the point (2, 1)
   Ollie says the image is the point (3, 2)
   Who is correct? Give a reason for your answer.

   b) Find the coordinates of the images of the other 2
   vertices of the triangle under this reflection.

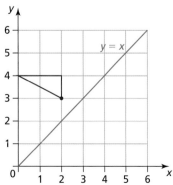

End of chapter reflection

You should know that...	You should be able to...	Such as...
A translation is described using a vector.	• Find the image of a point given the vector of a translation.	Find where the point (3, –2) moves to after a translation of $\begin{pmatrix} 3 \\ -4 \end{pmatrix}$
	• Solve problems involving vectors and translations.	The image of A(3, 1) under a translation is (–1, 4). Find the image of B(–3, –1) under the same translation.
A reflection on a coordinate grid is described by the equation of the mirror line.	• Reflect a shape on a coordinate grid using mirror lines such as $x = 3$, $y = -2$, $y = x$ and $y = -x$. • Identify the equation of a mirror line when given an object and an image.	Reflect the shaded shape: a) in the line $x = -1$ b) in the line $y = 1$
Shapes can be rotated on a coordinate grid around a centre of enlargement.	• Rotate shapes 90° and 180° around a centre of rotation on a coordinate grid.	Rotate shape A about the point (2, –1) by: a) 90° anticlockwise b) 180°

| Shapes can be enlarged on a coordinate grid using a centre of enlargement and a scale factor. | • On a coordinate grid, enlarge a shape with a given positive scale factor and centre of enlargement.

• Find the scale factor of an enlargement given the object and the image. | Shape T is enlarged.

State the scale factor.

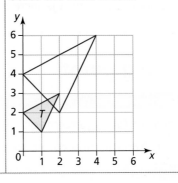 |

19 Percentages

You will learn how to:

- Understand percentage increase and decrease, and absolute change.

Starting point

Do you remember…

- how to find a percentage of a quantity, with and without a calculator?
  For example, find 80% of $240.

This will also be helpful when…

- you learn to understand and calculate compound percentages.

19.0 Getting started

A shop has a sale on all its items. Below shows the original price and the sale price for each item.

Pencils
Was $5.00
Now $4.00

Markers
Was $8.50
Now $6.80

Rubber
Was $0.95
Now $0.76

Was $2.75
Now $2.20

The shop has used a mathematical rule to reduce all of the prices. Can you work out what it is?

If the original price of an item in the shop was $4.00, find the new price.

19.1 Percentage increase and decrease

Key terms

An **absolute change** in a quantity is the amount by which the quantity increases or decreases. For example, if a price increases from $4.20 to $4.80, the absolute change is an increase of $0.60.

A **percentage change** in a quantity is a change in the quantity by a percentage of its original value.

A **percentage increase** makes a quantity bigger. For example, if a volume increases from 5 litres to 6 litres, that is a percentage increase of 20% (because the absolute increase, 1 litre, is 20% of 5 litres).

A **percentage decrease** makes a quantity smaller. For example, if a mass decreases from 24 kg to 6 kg, that is a percentage decrease of 75% (because the absolute decrease, 18 kg, is 75% of 24 kg).

A **multiplier** is a number by which you multiply a quantity to find its new value after a percentage change. For example, to increase a quantity by 20%, multiply it by 1.2. To decrease a quantity by 75%, multiply it by 0.25.

> **Did you know?**
>
> The percent symbol developed from the Italian words 'per cento', meaning per hundred. Over time the written words were shortened, until 'per' disappeared and 'cento' became two circles with a line between them: %.

Worked example 1

Increase 200 g by 16%

$100\% + 16\% = 116\%$ $= 1.16$ $200 \times 1.16 = 2 \times 116$ $= 232$ g	The original quantity is 100%, and you need to add another 16% To find 116% of 200 g, multiply by 1.16	**Alternative method** 10% of 200 = 20 1% of 200 = 2 6% of 200 = 6 × 2 = 12 16% of 200 = 20 + 12 = 32 200 g + 32 g = 232 g

Worked example 2

Decrease 260 m by 21%

$100\% - 21\% = 79\%$ $79\% = 0.79$ $260 \times 0.79 = 205.4$ m	Find the multiplier: to decrease by 21%, multiply by 0.79

1 Find:

a) 2% of 1200   b) 5% of 2300   c) 45% of 3000   d) 40% of 8600

e) 0.5% of 3000   f) 2.5% of 600   g) 300% of 800   h) 150% of 640

2 Find the absolute change when $400 is:

a) increased by 12%   b) decreased by 12%   c) increased by 400%

3 Write the multiplier for each calculation.

a) increase by 12%   b) decrease by 26%   c) increase by 3%   d) decrease by 80%

e) increase by 200%   f) decrease by 5%   g) increase by 132%   h) decrease by 99%

4 Increase:

a) $3500 by 1%   b) 1600 ml by 200%   c) 160 m by 10%   d) 240 kg by 25%

5 Decrease:

a) $5000 by 4%   b) 250 km by 5%   c) 180 m by 20%   d) 300 g by 25%

6 Use a calculator to find:

a) 13% of 350 kg   b) 45% of $82   c) 67% of 3200 km   d) 99% of 104 ml

e) 1.2% of $660   f) 500% of 19.5 cm   g) 120% of 962 ml   h) 22.5% of 56 kg

7 Use a calculator to find the answers. Use a multiplier for each calculation.

a) increase $3680 by 38%   b) decrease 3560 kg by 7%   c) increase 38.5 km by 49%

d) decrease 485 m by 38%   e) increase 660 ml by 225%   f) decrease 32 cm by 2.5%

g) increase 876 litres by 0.2%   h) decrease $450 by 97%   i) increase 178.5 cm by 48%

8 a) Maleshoane increases $100 by 100%. Her working is below.

$$100\% + 100\% = 200\%$$
$$200\% = 2$$
$$\$100 \times 2 = \$200$$

Is Maleshoane correct? Explain your answer.

b) Keme wants to increase $4.50 by 6%. His working is below.

$$1 + 0.6 = 1.6$$
$$\$4.50 \times 1.6 = \$7.20$$

Is Keme correct? Explain your answer.

9 In each pair of calculations, find the one with the larger amount.

a) Increase $60 by 20% or decrease $65 by 20%

b) Increase 400 kg by 40% or increase 420 kg by 35%

c) Increase 260 kg by 200% or decrease 600 kg by 15%

10 Kiana is asked to increase 200 ml by 50% and then increase the result by 50%.

She says, 'That is the same as increasing 200 ml by 100%. The result is 400 ml.'

a) Describe the mistake Kiana has made.

b) Find the correct answer.

11 Technology question The table shows the earnings of workers before any tax is taken off. The tax rate on earnings is 19%. How much does each worker earn after tax is taken off? Use a spreadsheet program to find how much each worker earns after tax is taken off.

Worker	Earnings before tax	Tax	Earnings after tax
a) Kira	$9800		
b) Shelley	$12 780		
c) Jas	$9500		
d) Ivan	$10 850		
e) Joshua	$22 200		

Think about

Decrease $200 by 100%

Is it possible to decrease a quantity by more than 100%?

Discuss

Do we need percentages? Couldn't we just use decimals (such as 0.47 instead of 47%) instead?

Thinking and working mathematically activity

Below are some quantities.

$15 $25 $40 $65 $100 $220

Describe what these quantities have in common.

Increase each quantity by 20%.

Describe and explain any properties shared by the resulting quantities.

Consolidation exercise   1–2

1 The prices of all train tickets increase by 10%. Find the new prices of these tickets.

RAIL LINE
CHILD $8
TRAVEL

RAIL LINE
ADULT $15
TRAVEL

2 Waqas buys a camera. Kasanita buys a printer.

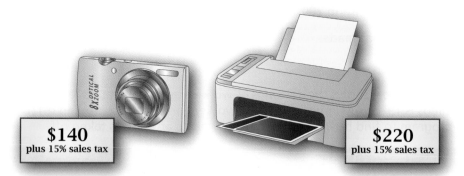

$140
plus 15% sales tax

$220
plus 15% sales tax

   a) Write a multiplier for working out the prices including tax.

   b) Find how much Waqas pays for the camera, including tax.

   c) Find how much Kasanita pays for the printer, including tax.

3 **a)** Sophie earns a salary of $32 300 per year. Her pay increases by 15%. Find her new salary.

   b) A school has 840 students. The number of students decreases by 5%.
      Find the new number of students.

4 Morten is buying a laptop. Show which deal gives the lower final price.

Was $545

NOW 34% off

Was $540

NOW $\frac{1}{3}$ off

5 Nur earns $19 500 per year with Company A. Company A offers her an 8% pay increase in her current job. Company B offers her a new job with a salary of $21 300.

With which company would Nur earn more?

6 State which gives the bigger absolute change in each pair.

   a) Increasing $91 by 21% or increasing $88 by 25%

   b) Increasing $256 by 37% or increasing $325 by 8%

   c) Increasing 58 ml by 100% or decreasing 232 ml by 100%

   d) Increasing 800 kg by 200% or decreasing 2000 kg by 25%

7 Branko wants to buy two books which have the same price.
There are two bookshops with special offers, as shown below.

HW Books

Book Sale
25% off

The Smiths

Buy one get
one half price

Is one offer better for Branko than the other? Explain your answer.

8 Decide whether each statement is true or false. Explain your answers.

   a) To increase a quantity by 350%, you use the multiplier 3.5.

   b) If you increase $500 by 10% and then decrease the result by 10%, you get $500.

   c) If you increase 60 kg by 100% and then decrease the result by 50%, you get 60 kg.

End of chapter reflection

You should know that...	You should be able to...	Such as...
You can use a multiplier to work out a percentage increase or decrease. The absolute change is the difference between the original amount and the new amount.	Increase or decrease a quantity by a percentage.	Increase $120 by 5% Decrease 850 km by 46%

You will learn how to:

- Understand term-to-term rules and generate sequences from numerical and spatial patterns (including fractions).
- Understand and describe nth term rules algebraically (in the form $n \pm a$, $a \times n$, or $an \pm b$, where a and b are positive or negative integers or fractions).

Starting point

Do you remember...

- how to continue simple integer sequences?

  For example, find the 7th term in the sequence that begins 11, 20, 29, 38, ...

- how to describe a simple integer sequence using a term-to-term rule?

  For example, find the term-to-term rule for the sequence –15, –3, 9, 21, ...

- how to describe the general term for spatial patterns?

  For example, what is the rule for finding the number of circles in any pattern in the sequence below?

  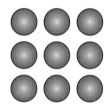 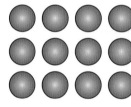

Pattern 1   Pattern 2   Pattern 3   Pattern 4

This will also be helpful when...

- you learn how to write an algebraic expression for any term in a sequence
- you find the inverse of a function.

20.0 Getting started

Eloa and Mancio receive money each birthday.

Eloa receives $7.50 on her first birthday.

The amount she receives increases by $1.50 each year.

On his 14th birthday Mancio receives $18. Each birthday, Mancio receives 50 cents more than the previous one.

- How much money does Eloa receive on each of her first five birthdays?
- Who receives the most money on their 10th birthday?

20.1 Generating sequences

> **Did you know?**
>
> Some sequences occur naturally. Search the internet to find out how sunflowers and pineapples are related to the Fibonacci sequence.

Worked example 1

a) The first term of a sequence is 4.7

   The term-to-term rule is **subtract 0.4**

   Find the 5th term.

b) The third term of a sequence is 1. The term-to-term rule is add $\frac{1}{4}$. Find the first term.

| a) The 5th term is 3.1 | The term-to-term rule tells you how to go from one term to the next. Here you subtract 0.4 every time. | 1st term   5th term
4.7   4.3   3.9   3.5   3.1
    −0.4   −0.4   −0.4   −0.4 |
| b) The second term is
$1 - \frac{1}{4} = \frac{3}{4}$
The first term is
$\frac{3}{4} - \frac{1}{4} = \frac{2}{4} = \frac{1}{2}$
The first term $= \frac{1}{2}$ | To get from one term to the next you add $\frac{1}{4}$. So to go backwards from the third term to the first term, you subtract $\frac{1}{4}$ each time. | 1st term   2nd term   3rd term
$-\frac{1}{4}$ $-\frac{1}{4}$
$\frac{1}{2}$ $\frac{3}{4}$ 1
$+\frac{1}{4}$ $+\frac{1}{4}$ |

Worked example 2

Here are the first four patterns in a sequence.

Pattern 1   Pattern 2   Pattern 3   Pattern 4

a) Write down the term-to-term rule for the sequence.

b) Find the number of circles in Pattern 7.

| a) The term-to-term rule is **add 4**. | Find the number of circles in each pattern. To go from one pattern to the next, 4 circles are added. | |

Pattern 1   1   Pattern 2   5   Pattern 3   9   Pattern 4   13

+4   +4   +4

| b) The number of circles in pattern 7 is 25. | The terms in the sequence increase in 4s, so to find the next term, add 4. | |

Pattern	1	2	3	4	5	6	7
Number of circles	1	5	9	13	17	21	25

Exercise 1

1 Write down the next three terms in each sequence.

a) 67, 78, 89, 100, …

b) 101, 95, 89, 83, …

c) −17, −13, −9, −5, …

d) 19, 14, 9, 4, …

e) 1.2, 1.5, 1.8, 2.1, …

f) 3.2, 2.8, 2.4, 2.0, …

g) $2\frac{1}{2}$, 4, $5\frac{1}{2}$, 7, …

h) $\frac{1}{7}$, $\frac{5}{7}$, $\frac{9}{7}$, $\frac{13}{7}$, …

2 Write down the term-to-term rule for each sequence and use it to find the 7th term each time.

a) 130, 118, 106, 94, …

b) 19, 28, 37, 46, …

c) 17, 11, 5, −1, …

d) −25, −18, −11, −4, …

e) 3.1, 3.6, 4.1, 4.6, …

f) $3\frac{1}{2}$, $3\frac{1}{4}$, 3, $2\frac{3}{4}$, …

3 Use the given information to find the required term in each of these sequences.

a) [1st term = 3.2, term-to-term rule = add 0.4]   Find the 4th term.

b) [1st term = 2.8, term-to-term rule = subtract 0.3]   Find the 5th term.

c) [1st term = $\frac{1}{10}$, term-to-term rule = add $\frac{3}{10}$]   Find the 5th term.

d) [4th term = 5.3, term-to-term rule = add 0.6]   Find the 1st term.

e) [8th term = −15, term-to-term rule = subtract 9]   Find the 3rd term.

4 Find the 5th term of each sequence:

a) 1st term = 6   term-to-term rule = 'multiply by 3'

b) 1st term = 40   term-to-term rule = 'divide by 2'

c) 1st term = 5   term-to-term rule = 'multiply by 2 and then subtract 4'

d) 1st term = 0.25   term-to-term rule = 'add 1 and then multiply by 4'

e) 1st term = −17   term-to-term rule = 'add 3 and then divide by 2'

f) 1st term = 69   term-to-term rule = 'multiply by $\frac{2}{3}$ and then subtract 4'

5 Technology question

a) Set up a spreadsheet with the number 2 in cells A1, B1 and C1.

Enter the formulae shown in cells A2, B2 and C2 and copy these formulae down to row 10.

◢	A	B	C
1	2	2	2
2	=A1*0.5+0.5	=B1*(−1)+1	=C1*4−6
3			
4			
5			

Think about

If you know the 6th term of a sequence, can you simply double it to get the 12th term?

b) Describe what is happening to the terms of the sequences generated in each of the columns.

6 The first number of a sequence is 10 and the term-to-term rule is 'multiply by 2 and subtract 5.'

a) How many terms of this sequence are less than 100?

b) Manish says that, after the first term, every term of this sequence has a units digit of 5. Explain why he is correct.

7 The first two terms of a sequence are 2 and 6. Complete the possible term-to-term rules for these two terms.

a) Multiply by ....

b) Multiply by 4 and 2

c) Add .... and multiply by 2

d) Multiply by 40 and ......

Thinking and working mathematically activity

Write down four sequences that have a second term of 11. The term-to-term rule must contain an addition or subtraction and multiplying or dividing.

Write down the term-to-term rule for each sequence.

Use the term-to-term rule to find the next two terms in each sequence.

---

8 The first term of a sequence is 2. The third term of the sequence is 50.

What could the second term be? Try to find four possible values, giving a reason for each of your answers.

9 Here are some of the patterns in a sequence made from squares.

a) Draw Pattern 3.

b) Find the term-to-term rule for the sequence.

c) Work out the number of squares needed to make Pattern 7.

  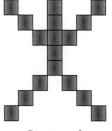

Pattern 1   Pattern 2   Pattern 3   Pattern 4

10 For each of these spatial patterns:

- Draw the next diagram.
- Find the number of squares in Pattern 7.
- Find the term-to-term rule.

a)

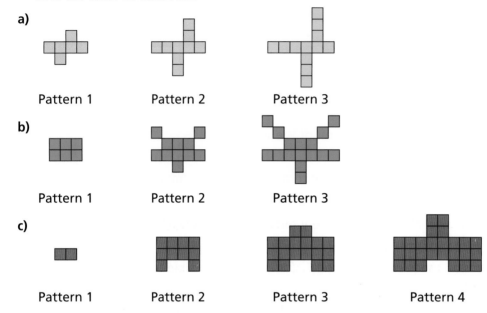

Pattern 1   Pattern 2   Pattern 3

b)

Pattern 1   Pattern 2   Pattern 3

c)

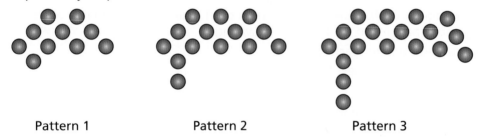

Pattern 1   Pattern 2   Pattern 3   Pattern 4

11 These three patterns are made from circular tiles.
The patterns are the first three terms in a sequence.

Explain why no pattern has an odd number of circular tiles.

Pattern 1   Pattern 2   Pattern 3

Key terms

A linear **sequence** is one that goes up or down by the same amount between terms. A linear sequence is also known as an arithmetic sequence.

It has a term-to-term rule of 'Add *a*' or 'Subtract *a*', where *a* is a fixed number.

You can write the **position-to-term rule** for a sequence in words, such as 'multiply the position by 2 and add 1', or we can write it as an algebraic expression called the **nth term**.

For example, instead of writing 'multiply the position by 2 and add 1', you could write $2n + 1$.

This expression shows *n* being multiplied by 2 and then 1 being added.

Worked example 3

a) Find the first three terms of the sequence with position-to-term rule 'multiply the position by 5 and then subtract 6'.

b) A second sequence has *n*th term given by $\frac{1}{2}n + 1$. Find the 8th term of the sequence.

c) A different sequence has *n*th term given by $6n - 7$. Show that the number 77 is a term in the sequence.

a) $1 \times 5 - 6 = -1$ $\quad 2 \times 5 - 6 = 4$ $\quad 3 \times 5 - 6 = 9$ The first three terms are -1, 4 and 9	To find the first three terms in the sequence, apply the rule using 1, 2 and 3 as inputs.	Position Term 1 -1 $2 \rightarrow \boxed{\times 5} \rightarrow \boxed{-6} \rightarrow 4$ 3 9
b) $\frac{1}{2}n + 1 = \frac{1}{2} \times 8 + 1$ $\qquad\qquad = 4 + 1$ $\qquad\qquad = 5$	Substitute $n = 8$ into the *n*th term rule.	The position-to-term rule for this sequence is 'multiply the position by $\frac{1}{2}$ and then add 1'. Use 8 as the input. Position Term $8 \rightarrow \boxed{\times \frac{1}{2}} \xrightarrow{4} \boxed{+1} \rightarrow 5$
c) $6n - 7 = 77$ $\quad\; 6n = 84$ $\quad\;\; n = 84 \div 6$ $\qquad\; = 14$ So 77 is a term in the sequence. It is the 14th term.	Look to see if there is a position value that gives 77 as a term in the sequence. Solve $6n - 7 = 77$. If the solution is a whole number then 77 is a term in the sequence.	Position Term $n \rightarrow \boxed{\times 6} \rightarrow \boxed{-7} \rightarrow 77$ $14 \leftarrow \boxed{\div 6} \leftarrow \boxed{+7} \leftarrow 77$ 84

Why is a position-to-term rule more useful than a term-to-term rule?

Worked example 4

Find an expression for the nth term of each sequence.

a) 0.7, 1.4, 2.1, 2.8, ….

b) –2, 1, 4, 7, ….

c) 6, 2, –2, –6, …

a) The nth term is $0.7n$	0.7 1.4 2.1 2.8 + 0.7 + 0.7 + 0.7 The terms in the sequence are multiples of 0.7	$0.7 \times 1 = 0.7$ $0.7 \times 2 = 1.4$ $0.7 \times 3 = 2.1$ $0.7 \times n = 0.7n$
b) The nth term is $3n - 5$	The term-to-term rule for the sequence is 'add 3'. So compare the terms of the sequence with the multiples of 3. The terms of the sequence are 5 less than the multiples of 3.	
c) The nth term is $-4n + 10$ This could also be written as $10 - 4n$	The term-to-term rule for the sequence is 'subtract 4'. Compare the terms in the sequence with the multiples of –4. The terms of the sequence are 10 more than the multiples of –4.	

Exercise 2

1 The rule for generating a sequence is: term in sequence = 3 × position number + 7

Find the terms that are in the following positions:

a) position 1   b) position 4   c) position 10

2 Find the sixth term in the sequences generated by these nth term rules:

a) $3n + 11$ b) $7n - 10$ c) $30 - 4n$ d) $\dfrac{n}{2} + 5$

3 A sequence has nth term $12 - \dfrac{1}{3}n$. Which of these statements is true? Explain your answer.

a) The 6th term = 9   b) The 12th term = 0   c) The 15th term = 7   d) The 36th term = 0

4 Match each *n*th term to the sequence it describes.

3*n* + 1	7, 9, 11, 13, …
3*n*	2, 4, 6, 8, …
2*n*	7, 11, 15, 19, …
2*n* + 5	3, 6, 9, 12, …
4*n*	4, 7, 10, 13, …
5*n* − 2	4, 8, 12, 16, …
4*n* + 3	3, 8, 13, 18, …

5 Here are the *n*th terms of some sequences.

Make a copy of the table below and then write each sequence in the correct column.

Sequence A	Sequence B	Sequence C	Sequence D
4*n* − 2	2*n* − 4	−2*n* + 4	2*n* + 4

Sequence E	Sequence F	Sequence G	Sequence H
−4*n* + 4	−2*n* + 2	4*n* + 2	−4*n* + 2

Terms go up in 2s	Terms go up in 4s	Terms go down in 2s	Terms go down in 4s

6 The *n*th term of a sequence is 7*n* + 9.

Find whether or not each of the following numbers are terms in the sequence.
Give reasons for your answers.

a) 85 **b)** 65 **c)** 121 **d)** 150

7 a) A sequence has *n*th term $4 - \frac{1}{2}n$. Mary thinks that the 8th term of the sequence is the first

term that is negative. Show that Mary is wrong.

b) The position-to-term rule for a sequence is 'multiply by 3.5 and add 11'. Find the first term that is greater than 100.

8 Here is some information about three sequences.

Juan's sequence	Laila's sequence	Dominik's sequence
The position-to-term rule is 'multiply by 4 and add 3'.	The position-to-term rule is 'multiply by 6 and subtract 10'.	The term-to-term rule is 'subtract 6'.

a) Find the 20th term in Juan's sequence.

b) Laila says, 'The 6th term in my sequence is greater than the 6th term in Juan's sequence.'

Show that Laila is wrong.

c) The 4th term in Laila's sequence is equal to the 4th term in Dominik's sequence.

Find the 1st term in Dominik's sequence.

9 Find how each of these sequences is related to the number in the 7 times table.

a) 8, 15, 22, 29, …   b) 17, 24, 31, 38, …   c) 5, 12, 19, 26, …   d) 1, 8, 15, 22,

10 Complete the nth term rules for these sequences:

Sequence	nth term rule
a) 11, 15, 19, 23, …	$4n + $ …..
b) 13, 22, 31, 40, …	$9n + $ …..
c) 5, 13, 21, 29, …	$8n - $ …..
d) 8, 15, 22, 29, …	….. $n + 1$
e) 11, 16, 21, 26, …	….. $n + 6$
f) 9, 20, 31, 42, …	….. $n - 2$

11 Find the nth term for each sequence:

a) 6, 9, 12, 15, …   b) 10, 18, 26, 34, …   c) 9, 21, 33, 45, …   d) −1, 5, 11, 17, …

e) 3, 12, 21, 30, …   f) 29, 45, 61, 77, …   g) −12, −10, −8, −6, …   h) 1.5, 2, 2.5, 3, …

12 Find the nth term for these sequences:

a) 19, 15, 11, 7, …   b) 15, 8, 1, −6, …   c) 80, 72, 64, 56, …

d) $\frac{4}{5}$, $1\frac{4}{5}$, $2\frac{4}{5}$, $3\frac{4}{5}$, …   e) 1.3, 2.6, 3.9, 5.2, …   f) $-\frac{1}{2}$, −1, $-\frac{3}{2}$, −2, …

g) 0.7, 1.8, 2.9, 4, …   h) $\frac{1}{10}$, $\frac{3}{10}$, $\frac{5}{10}$, $\frac{7}{10}$, …

13 Nik collects toy cars. At the start of week 1, Nik has 46 cars. He buys 3 cars each week.

a) Find how many cars Nik will have at the start of week 5.

b) Find an expression for how many cars Nik will have at the start of week n.

c) Nik says that he will have more than 200 cars in his collection at the start of week 50.

Is he correct? Show how you worked out your answer.

14 Write an expression for the number of squares in the *n*th pattern of each sequence.

a)

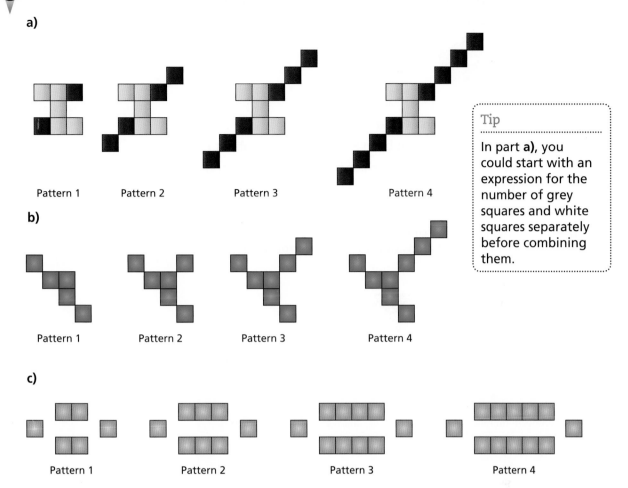

Pattern 1   Pattern 2   Pattern 3   Pattern 4

> **Tip**
>
> In part **a)**, you could start with an expression for the number of grey squares and white squares separately before combining them.

b)

Pattern 1   Pattern 2   Pattern 3   Pattern 4

c)

Pattern 1   Pattern 2   Pattern 3   Pattern 4

> **Think about**
>
> Does the *n*th term rule '3 × *n* + 4' give a different sequence to the rule '(*n* + 4) × 3'?

15 The third term of a sequence is 14. The term-to-term rule is 'add 6'.

a) Find the *n*th term.

b) Find the 100th term

> **Discuss**
>
> Which sequence is the odd one out? Give a reason for your choice.
>
> 13, 16, 19, 22, …
>
> 13, 15, 17, 19, ….
>
> 5.3, 8.3, 11.3, 14.3, …
>
> 4.2, 7.2, 10.2, 13.2,…

Thinking and working mathematically activity

Here is a pattern of hexagons:

Pattern 1   Pattern 2   Pattern 3

- Can you draw Pattern 4 and Pattern 5?

- How many hexagons will there be in Pattern 20? What about Pattern *n*?

- Now calculate the perimeter of each of the patterns 1 to 5 and record your results in a table.

- What do you notice?

- Can you find a position-to-term rule for the perimeter of Pattern *n*?

- Investigate the perimeter of the shapes made by the white space enclosed by the shaded hexagons – can you find a position-to-term rule for this too?

Consolidation exercise

 1 For each sequence, decide if the terms are always odd, always even or sometimes odd and sometimes even.

A	B
First term = 18 Term-to-term rule is 'Add 7'	First term = 25 Term-to-term rule is 'Add 4'

C	D
Position-to-term rule of sequence is 'Multiply by 6 and add 14'	Position-to-term rule of sequence is 'Multiply by 11 and add 5'

2 Match each term to the sequence it comes from.

A	P
1st term = 5.5 Term-to-term rule is 'Subtract 0.7'	5th term = 2.7

B	Q
1st term = 2.4 Term-to-term rule is 'Add 0.6'	5th term = 4.6

C	R
*n*th term rule is '3*n* − 10.4'	5th term = 4.8

D	S
*n*th term rule is '0.8*n* + 1.2'	5th term = 5.2

3 Here is a sequence of patterns made from circles.

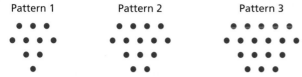

Pattern 1   Pattern 2   Pattern 3

a) Find a rule for finding the number of circles in any pattern.

b) Rania says that 86 circles are needed for Pattern 20. Is she correct?
Show how you found your answer.

4 Aroush and Harry have different rules for generating sequences.

Aroush: Position-to-term rule: 'divide by 2 and subtract 18.'

Harry: First term = 27, term-to-term rule is 'subtract 3'.

What is the position of the term in Harry's sequence which is equal
to the 12th term in Aroush's sequence?

5 Here is a sequence of patterns.

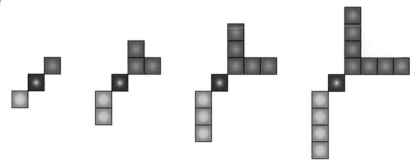

Write an expression for the total number of squares in the nth pattern.

6 Sabina is exploring a sequence. Here are the first five terms:

10, 18, 26, 34, 42, …

She says, 'My sequence is just 2 more than the 8 times table'.

a) Do you agree with Sabina? Explain your answer.

b) Write an algebraic expression for the nth term of the 8 times table.

c) Now write an algebraic expression for the nth term of Sabina's sequence. What do you notice?

7 The position-to-term rule for a sequence is 'multiply by 2.5 and add 1.5'.
Which of these numbers are terms in the sequence?

   1.5   2.5   4   9   18   21.5

8 Match each pattern to the expression for its *n*th term and its number sequence. Complete the missing pattern, expression and number sequence to form 5 matching pairs.

	2*n* + 3	4, 7, 10, 13, …
	2*n* + 1	5, 9, 13, 17, …
		7, 11, 15, 19, …
	3*n* + 1	
	4*n* + 3	5, 7, 9, 11, …

9 a) Match each sequence to its *n*th term rule.

   5, 8, 11, 14, …
   −1, 1, 3, 5, …   3*n* − 2   2*n* − 3   3*n* + 2   2*n* + 3   3*n* + 3
   1, 4, 7, 10, …
   6, 9, 12, 15, …

b) Find the sequence that is generated from the *n*th term rule that does not have a pair.

10 A teacher asks three students to think of a sequence that contains the number 20.
The *n*th term rules of their sequence are:

Nadia's sequence	Marco's sequence	Tania's sequence
7*n* − 1	32 − 3*n*	3*n* + 1

Which of the students have found a sequence that contains the number 20? Explain your answer.

End of chapter reflection

You should know that...	You should be able to...	Such as...
A linear (or arithmetic) sequence can be defined by a term-to-term rule. An arithmetic sequence is one with a term-to-term rule of $+a$ or $-a$, where a is a constant number.	Use and find a term-to-term rule of a linear sequence. Use term-to-term rules in connection with spatial patterns.	Find the 8th term of the sequence that begins 16, 10, 4, ... Find the number of dots of the next pattern in this sequence:
An nth term rule can be used to generate a sequence.	Use a given nth term rule to find a term in the sequence. Find an expression for the nth term of a linear sequence.	a) The nth term for a sequence is given by the expression $4n - 15$. Find the 10th term. b) Find the nth term for the linear sequence that begins 14, 23, 32, 41, ...

Probability 2

You will learn how to:

- Understand that tables, diagrams and lists can be used to identify all mutually exclusive outcomes of combined events (independent events only).
- Understand how to find the theoretical probabilities of equally likely combined events.
- Record, organise and represent categorical, discrete and continuous data.
  - ○ Venn diagrams

Starting point

Do you remember…

- how to list the outcomes for a single event?

  For example, list all of the possible outcomes when you roll a fair six-sided dice.

- how to calculate probabilities for a single event?

  For example, find the probability of rolling a two on a fair six-sided dice.

This will also be helpful when…

- you learn how to draw two-way tables, sample space diagrams and tree diagrams to show all possible outcomes of two or more successive events.

21.0 Getting started

Use three cards with a triangle and two circles printed on them.
Turn them over so that only you know which card the triangle is on.

Ask your partner to guess which card the triangle is on, but do not yet reveal it.

Instead, reveal a card which has a circle on it.

Ask your partner if they wish to stick with their original choice or switch to the other remaining card.
They win if they select the triangle.

Repeat the game several times and record the results.

Do they win more games if they switch or stick?

What are the outcomes for this game?

Are both outcomes equally likely?

What is the probability that the first card selected is a triangle?

21.1 Lists and tree diagrams for combined events

Key terms

The **event** of flipping a coin has two **outcomes**: heads or tails.

Two outcomes are **mutually exclusive** if they cannot happen at the same time.
For example, you cannot roll a four and a six on a single roll of a six-sided dice.

When you work through a problem **systematically** you should follow a logical order to ensure that nothing gets missed. For example, when flipping a coin twice, list all the possibilities of getting a head first before moving on to list all the possibilities of getting a tail first.

Successive events occur one after the other or at the same time.
For example, you flip one coin twice or you roll two dice at the same time.

Tree diagrams are used to display **combinations** of two or more events and are a way to make sure no events are missed out.

Did you know?

In a room of just 30 people, there is a 70% chance that at least two people will share the same birthday. This may seem like it wouldn't be possible, but it is true, and it is called The Birthday Paradox.

Worked example 1

There are two bags of sweets at a party. Mitchel takes one sweet from each bag at random.

Bag A

Bag B

a) List all the possible combinations of two sweets.

b) Find the probability that Mitchel takes two sweets of the same colour.

c) Find the probability that Mitchel does not take a blue or a green sweet.

a) The outcomes are:

red, red

red, blue

red, yellow

blue, red

blue, blue

blue, yellow

yellow, red

yellow, blue

yellow, yellow

green, red

green, blue

green, yellow

There are 12 combinations.

It can be useful to draw a tree diagram to organise the outcomes of two or more events systematically and make sure that none are missed out.

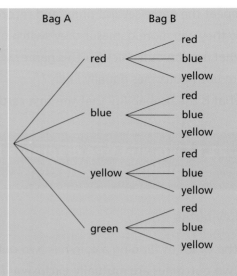

b) $\frac{3}{12}$ or $\frac{1}{4}$

It can be useful to draw a table so you remember which bag of sweets you are thinking about.

You can write your list in any order, but it is a good idea to do it systematically. You could first list all possibilities when taking a red from the first bag, and follow the same order (red, blue, yellow) with the second bag.

There are 3 out of 12 outcomes which have the same colour sweet being chosen.

Bag A	Bag B
red	red
red	blue
red	yellow
blue	red
blue	blue
blue	yellow
yellow	red
yellow	blue
yellow	yellow
green	red
green	blue
green	yellow

Three of the rows have the same colour sweet.

c) $\frac{4}{12}$ or $\frac{1}{3}$

There are 4 out of 12 outcomes that do not have a blue or a green sweet being chosen.

Bag A	Bag B
red	red
red	blue
red	yellow
blue	red
blue	blue
blue	yellow
yellow	red
yellow	blue
yellow	yellow
green	red
green	blue
green	yellow

Four of the rows do not include a blue or green sweet.

Exercise 1

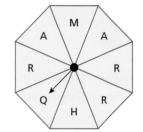

1 Here is a spinner.

List all the possible outcomes when the spinner is spun once.

2 Henri has three cards, numbered 2, 5 and 8. He wants to find how many three-digit numbers he can make using the cards. He starts to draw a table to organise the results.

a) Copy and complete Henri's table. Write down how many three-digit numbers he can make.

Digit 1	Digit 2	Digit 3
2	5	8
2		

b) One of the three-digit numbers is selected at random. What is the probability that it is an even number?

3 A game is played by spinning each of these spinners once.

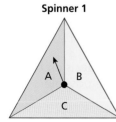

Spinner 1

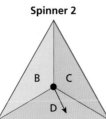

Spinner 2

a) Copy and complete the table to show all the possible outcomes.

Spinner 1	Spinner 2
A	B
A	C

You score a point by getting the same letter on both spinners.

b) What is the probability that you score a point on your turn?

4 A restaurant has this menu.

with

Bella makes a choice from the menu at random.

a) Draw a tree diagram to show all the possible outcomes.

b) Find the probability that Bella has quiche and a jacket potato.

c) Find the probability that she has noodles or salad (or both).

5 There are four doughnuts in a bag.

Two are jam filled.

One is chocolate filled.

One is cream filled.

Augustus picks out a doughnut and eats it. He then takes another doughnut.

a) Copy and complete the table to show all possible combinations.

b) What is the probability that Augustus takes two jam doughnuts?

c) What is the probability that he takes a cream doughnut?

d) What is the probability that he takes two chocolate doughnuts?

1st doughnut	2nd doughnut
jam	jam
jam	chocolate

> **Discuss**
>
> A coin is flipped twice.
>
> Is the probability of obtaining two heads $\frac{1}{3}$ or $\frac{1}{4}$?
> Explain your decision by listing all the possible outcomes.

6 Lucy has a fair four-sided dice with sides labelled: 1, 2, 3 and 4.

She rolls the dice twice and adds the scores together.

a) Copy and complete the list to show all the possible outcomes.

(1, 1), (1, 2), (1, 3), ...

b) What is the probability that she scores a total of 5?

c) What is the probability that she gets a total greater than 6?

7 Kim wants to write a letter.

She can use white, blue or yellow paper.

She puts her letter in either a white, red or blue envelope.

Think about

How many possible outcomes are there when you roll two ordinary dice? Would a tree diagram be a suitable way to display all the outcomes?

a) Draw a tree diagram to show all the possible combinations of colours for paper and envelope.

b) Kim chooses the colour of the paper and the colour of the envelope at random. Find the probability that the colours of paper and envelope are the same.

8 A restaurant serves a choice of the two starters and three main meals as shown on the menu.

Mikael says that there are 9 possible combinations of choosing a starter and a main meal and he writes them all down.

a) Complete the tree diagram.

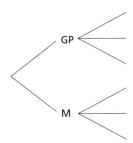

b) Explain what mistake Mikael has made when he listed his outcomes.

9 Aeeza has three children.

Find the probability that she has three girls.

Tip

Write a list of all the possible outcomes, starting with: Boy Boy Boy (BBB), Boy Boy Girl (BBG) etc.

Thinking and working mathematically activity

A menu has 3 different starters and 4 different main courses. A person can choose a starter and a main course.

How many different combinations are possible?

Investigate how the number of meal combinations changes as the number of starters and main courses changes.

A different restaurant offers a choice of starters, main courses and desserts. There are 36 possible menu combinations if someone has a starter, main course and dessert.

How many starters, main courses and desserts could there be? Convince a friend your solution is correct.

21.2 Sample space diagrams and Venn diagrams

Key terms

When this spinner is spun twice, there are nine equally likely outcomes:

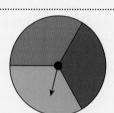

Red, Red	Red, Blue	Red, Orange
Blue, Red	Blue, Blue	Blue, Orange
Orange, Red	Orange, Blue	Orange, Orange

Instead of writing all the outcomes for two successive events as a list, they can be shown as a diagram. **Sample space diagrams** (or **possibility diagrams**) are a useful way to summarise information.

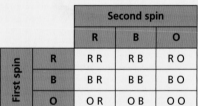

		Second spin		
		R	B	O
First spin	R	R R	R B	R O
	B	B R	B B	B O
	O	O R	O B	O O

The equally likely outcomes can be sorted into a **Venn diagram**.

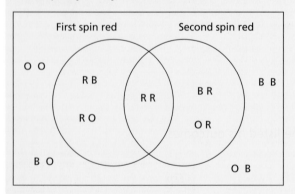

Worked example 2

Amy and Carlo each have a bag containing numbered counters.

Amy   Carlo

 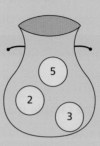

They each pick a counter from their bag at random.

They add together the numbers on their counters to get a total.

a) Draw a sample space diagram to show the possible **totals**.

b) Find the probability that the sum of their numbers is an even number.

a)

Carlo's bag

	2	3	5
1	3	4	6
2	4	5	7
2	4	5	7
4	6	7	9

Amy's bag

Draw the sample space diagram. Amy's bag contains four green counters numbered 1, 2, 2 and 4. Carlo's bag contains 3 yellow counters numbered 2, 3 and 5.

Carlo's bag

	2	3	5
1			
2			
2			
4			

Amy's bag

Add the different combinations of counters to complete the sample space diagram. For example, if Amy selects a counter numbered 4 and Carlo selects a counter numbered 5, this totals to 9.

Carlo's bag

	2	3	5
1	3	4	6
2	4	5	7
2	4	5	7
4	6	7	9

Amy's bag

b) P(even number) = $\frac{5}{12}$

There are 12 possible outcomes shown in the sample space diagram. Of these, 5 are even.

probability =

$$\frac{number\ of\ favourable\ outcomes}{total\ outcomes}$$

The probability that the total is an even number is $\frac{5}{12}$

Carlo's bag

	2	3	5
1	3	4	6
2	4	5	7
2	4	5	7
4	6	7	9

Amy's bag

1 A coin is tossed and an ordinary six-sided dice is thrown.

a) Copy and complete the sample space diagram for the possible outcomes of the two events.

	1	2	3	4	5	6
Heads, H	H, 1					H, 6
Tails, T			T, 3			

b) How many entries are in the sample space diagram?

2 The spinner shown is spun twice.

a) Complete the sample space diagram to show all the possible outcomes.

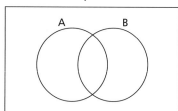

		Second spin			
		1	**2**	**3**	**4**
First spin	**1**	1, 1	1, 2	1, 3	1, 4
	2	2, 1			
	3	3, 1			
	4	4, 1			

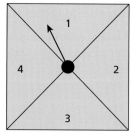

b) A is the outcome 'the total of the two scores is 5'. B is the outcome 'the first spin lands on 2'.
Write the 16 possible outcomes in the correct position on a copy of this Venn diagram.

3 A spinner is spun and can land on three possible colours – red (R), blue (B) and white (W).
Then a spinner, which has four sectors labelled 1,2, 3 and 4 respectively, is spun.

a) Draw the sample space diagram for the possible outcomes of the two events.

b) How many entries are in the sample space diagram?

4 Two fair dice, each with 6 faces labelled 1, 2, 3, 4, 5 and 6 respectively,
are thrown and the numbers shown on the two dice are added.

a) Copy and complete the sample space diagram for this event.

		First dice					
		1	**2**	**3**	**4**	**5**	**6**
Second dice	**1**	2		4			
	2						
	3				7		
	4						
	5		7				
	6						12

b) Which number occurs most often in the sample space diagram?

c) How many entries are there in the sample space diagram?

d) What is the probability that the sum of the two numbers shown on the dice is 4?

e) What is the probability that the sum of the two numbers shown on the dice is 7?

f) What is the probability that the sum of the two numbers shown on the dice is less than 4?

5 Jani has two bags, *A* and *B*, each containing four numbered counters.

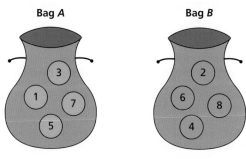

Bag A **Bag B**

Jani takes a counter at random from bag *A* and a counter at random from bag *B*.
He multiplies the numbers on the two counters to get a score.

a) Copy and complete the table to show all the possible scores.

			Bag A		
	×	1	3	5	7
Bag B	2	2	6		
	4				
	6			30	
	8				

b) Jani says the probability of getting a score of 6 is $\frac{1}{8}$. Is he correct? Explain your reasoning.

6 **a)** List all the three-digit numbers that can be made with the digits 2, 5 and 7.

b) A is the event 'even number'. B is the event 'middle digit is 5'.

Copy the Venn diagram and write each three-digit number from your list in the correct position.

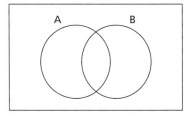

7 Two fair four-sided spinners, each labelled 1, 2, 3 and 4 respectively, are spun.
The numbers shown on the spinners are multiplied together.

a) Make a sample space diagram to show the possible outcomes.

b) How many entries are in the sample space diagram?

c) Which number occurs most often in the sample space diagram?

d) Find the probability that the product of the two numbers shown is 4.

e) Find the probability that the product of the two numbers shown is 3.

8 Two fair, six-sided dice are thrown and the numbers shown on the two dice are added.

Sura says that because there are 11 possible results of adding the two numbers shown (2, 3, 4, 5, 6, 7, 8, 9, 10, 11 and 12) then the probability that the sum of the two numbers is 12 is $\frac{1}{11}$. Explain why this is not the case.

9 A spinner with n sides is spun and the number it lands on is noted. The same spinner is then spun again, and the number shown the second time is added to the number shown the first time. A sample space diagram is drawn to show the possible outcomes. How many entries are in the sample space diagram?

Thinking and working mathematically activity

Ellie and Ahmed are playing a dice game. They throw two ordinary dice and then find the difference between the scores on each dice.

Ellie wins if the difference is less than 2.

Ahmed wins if the difference is 2 or more.

Ahmed says this is a fair game. Is he correct? Give reasons for your answer.

Suggest a way to change the game so that they have equal probabilities of winning.

Consolidation exercise

1 The only possible outcomes when you spin this 6-sided spinner are X, Y and Z.

Complete the spinner so that X is twice as likely to occur as Y.

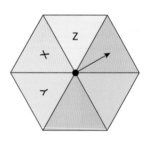

2 The tree diagram shows all the possible outcomes when you spin the two spinners below right.

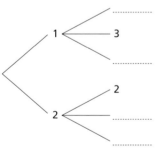

Spinner A   Spinner B

1 ⟨ 3

2 ⟨ 2

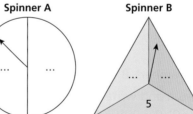

Spinner A   Spinner B

a) Complete a copy of the tree diagram and fill in the missing numbers on the spinners.

b) Event P is a total of 4 on the spinners. Event Q is spinner B landing on 2.

Copy and complete the Venn diagram to show all the possible outcomes.

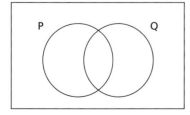

P   Q

---

3 These four cards are placed in a bag.

One card is selected and then put back in the bag. Another card is then selected.

a) List all the possible outcomes. The first one is done for you.
(triangle, triangle), ...

b) Match the statements with the correct probability:

A triangle and a rectangle are selected. $\frac{1}{16}$

At least one trapezium is seleted. $\frac{3}{4}$

Two circles are selected. $\frac{1}{8}$

Two different shapes are selected. $\frac{7}{16}$

4 A four-sided dice is rolled and a fair coin is spun.

a) Complete the sample space diagram to show all possible outcomes.

	Dice			
	1	**2**	**3**	**4**
Coin **H**	H,1	H,2		
T				

b) State whether each statement below is true or false.

i) There are 6 possible outcomes.

ii) The probability of getting a head and an even number is $\frac{1}{4}$.

iii) The probability of getting a tail and a 3 is $\frac{1}{8}$.

iv) The probability of getting an odd number is $\frac{3}{8}$.

5 Siobhan spins the spinner once.

She says, 'The probability of getting an A and a B is $\frac{5}{8}$.'

Is Siobhan correct? Give a reason for your answer.

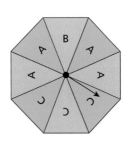

6 Hayley has to cross the city centre to get to work. The options for the two stages of her journey to work are shown in the table below.

Stage 1	Stage 2
walk	walk
bus	bus
tram	taxi

a) Copy and complete the tree diagram.

Stage 1   Stage 2

walk   walk

b) Hayley chooses the travel option for each stage at random. Find the probability that she walks all the way to work.

c) What is the probability that she does not get a bus at either stage of her journey?

7 The sample space diagram for the outcomes when two spinners, A and B, are spun, has 24 entries. Each spinner has at least 3 sectors. How many sides could spinner A have? all the possible answers

End of chapter reflection

You should know that...	You should be able to...	Such as...
Mutually exclusive events cannot happen at the same time.	Recognise which outcomes are mutually exclusive. List all the mutually exclusive outcomes of a single event.	Which of the following outcomes are mutually exclusive? **a)** Getting a 3 and a 4 when you roll two dice. **b)** Choosing 'banana' and 'mint choc chip' when selecting one scoop of ice-cream. **c)** Write down all the possible outcomes when you roll an ordinary six-sided dice.
The outcomes of two successive events are not mutually exclusive.	List all of the outcomes of two successive events. Calculate probabilities of two successive events by considering all of the possible outcomes.	This spinner is spun twice and the scores are added together. **a)** List all of the possible outcomes. **b)** Find the probability of getting a total of 9.

A tree diagram can be drawn to show the possible outcomes of two or more events.	Draw a tree diagram from information about the events and their outcomes.	Complete the tree diagram to show all the possible outcomes when a coin is flipped twice.
		First flip Second flip H .... T ....
A sample space diagram can be drawn to show the possible set of outcomes of two events.	Draw a sample space diagram from information about the events and their outcomes.	This spinner is spun twice. Draw a sample space diagram to show the possible outcomes. Find the probability of the spinner landing on a red sector on both spins.
Equally likely outcomes can be sorted into Venn diagrams.	Complete a Venn diagram from information about the events and their outcomes.	A four-sided dice is rolled twice. Event A is an even total when the two scores are added together. Event B is a 3 on the second roll. Complete the Venn diagram to show the possible outcomes. A B

You will learn how to:
- Understand and use the relationship between ratio and direct proportion.
- Use knowledge of equivalence to simplify and compare ratios (different units).
- Understand how ratios are used to compare quantities to divide an amount into a given ratio with two or more parts.

Starting point

Do you remember…

- how to simplify and compare ratios where both sides have the same units?

  For example, write each ratio in its simplest form: (a) 80 cm : 48 cm (b) 2 kg : 1.5 kg

- how to use the unitary method to solve problems involving ratio?

  For example, if 8 identical bags of flour have mass 6400 g, find the mass of 3 bags of flour.

- how to divide a quantity into two parts using a ratio?

  For example, share $90 in the ratio 8 : 7

This will also be helpful when…

- you learn to solve more difficult problems involving ratio
- you learn about the relationship between two quantities that are in direct proportion.

22.0 Getting started

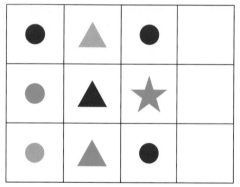

Three shapes are missing from the last column in the diagram above. Use the clues below to find the missing shapes and their colours.

- There are no blue stars.
- The fraction of shapes that are circles is $\frac{1}{2}$
- The ratio of green shapes to orange shapes is 2 : 1
- The fraction of shapes that are orange is $\frac{1}{6}$
- The ratio of stars to circles is 1 : 3
- The fraction of shapes that are blue is $\frac{1}{2}$

22.1 Relating ratio and proportion

Key terms

A **proportion** shows the size of a part relative to the whole. It is usually expressed as a fraction.

Worked example 1

a) In a bag of red buttons and yellow buttons, the proportion of red buttons is $\frac{5}{7}$

Find the ratio of red buttons to yellow buttons.

b) In a bag of blue buttons and green buttons, the ratio of blue buttons to green buttons is 1 : 2

Write the proportion of blue buttons.

a) Proportion of yellow buttons: $$1 - \frac{5}{7} = \frac{2}{7}$$	If $\frac{5}{7}$ of the buttons are red, the rest must be yellow.
Ratio of red to yellow : $\frac{5}{7} : \frac{2}{7} = 5 : 2$	To simplify the ratio, multiply both sides by 7.
b) $\frac{1}{3}$	The number of parts is 1 + 2 = 3 One out of these, three parts is green.

▼ Thinking and working mathematically activity

Make each set of red and yellow counters as described below.

For each set, write the proportion of yellow counters, and the ratio of red counters to yellow counters.

- a set where the proportion of red counters is $\frac{1}{3}$
- a different set where the proportion of red counters is $\frac{1}{3}$
- a set where the proportion of red counters is $\frac{3}{5}$
- a different set where the proportion of red counters is $\frac{3}{5}$

Describe and explain the relationship between the proportions of red and yellow counters, and the ratio of red to yellow counters.

Exercise 1

1 Alan and Long share 15 marbles. Alan gets 7 and Long gets 8.

a) Write the ratio of Alan's marbles to Long's marbles.

b) Write the ratio of Long's marbles to Alan's marbles.

c) What fraction of the marbles does Alan get?

d) What fraction of the marbles does Long get?

2 Thiemta and Suu share 60 sweets. Thiemta gets 36 sweets.

 a) Write the ratio of Thiemta's sweets to Suu's sweets in its simplest form.

 b) What fraction of the sweets does Thiemta get? Write the fraction in its simplest form.

 c) What fraction of the sweets does Suu get? Write the fraction in its simplest form.

3 The table shows information about girls and boys in four schools. Copy the table and fill in the gaps.

School	Ratio of girls to boys	Proportion of girls	Proportion of boys
A	1 : 2		
B	4 : 3		
C		$\frac{3}{5}$	
D			$\frac{1}{2}$

4 Blue and red paint are mixed to make purple paint. The proportion of blue paint is $\frac{2}{5}$.

 a) Write the proportion of red paint, as a fraction.

 b) Write the ratio of blue paint to red paint.

 c) How much blue and red paint are needed to make 20 litres of purple paint?

5 Mia and Hassan hire a car. Hassan pays $\frac{2}{3}$ of the bill.

 a) What proportion does Mia pay? Write this as a fraction.

 b) Write the ratio of Mia's payment to Hassan's payment.

 c) If the cost of the hire care is $240, find how much they each pay.

6 300 g of sugar is mixed with 750 g of flour.

 a) What proportion of the mix is sugar? Write this as a fraction in its simplest form.

 b) What proportion of the mix is flour? Write this as a fraction in its simplest form.

 c) Write the ratio of sugar to flour in its simplest form.

7 In a school there are 60 adults who represent $\frac{1}{10}$ of the population.
The rest of the school's population are students.

 a) What proportion of the school's population is students?

 b) Write the ratio of adults to students.

 c) Find the number of students in the school.

8 Ali and Bill share some money. Ali receives $\frac{2}{5}$ of the money.

 Bill says, 'Ali and I share the money in the ratio 2 : 5'.

 Is Bill correct? Explain your answer.

22.2 Simplifying and comparing ratios

Key terms

Equivalent ratios are ratios that show the same relationship.

For example, the ratios of oranges to lemons to limes 1 : 2 : 3 and 2 : 4 : 6 are equivalent ratios. 2 : 4 : 6 shows that there are 2 oranges for every 4 lemons and 6 limes, which is the same relationship as 1 lemon for every 2 oranges and 3 limes.

The **simplest form** of a ratio is the equivalent ratio that has the smallest possible whole numbers.

For example, the simplest form of the ratio 4.5 : 3 : 9 is 3 : 2 : 6

Did you know?

If the ratio of the lengths of a triangle's sides is 3 : 4 : 5, the triangle is right-angled.

You can make a right angle by tying knots to divide a rope into 12 equal parts. Then use the rope to form a triangle with sides in the ratio 3 : 4 : 5. The angle between the two shorter sides will be 90°.

Ancient Egyptian pyramid builders may have used this method to make right angles.

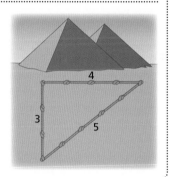

Worked example 2

a) Write the ratio 0.2 : 0.4 : 0.6 in its simplest form.

b) Write the ratio $\frac{3}{4} : \frac{1}{8} : \frac{1}{4}$ in its simplest form.

a) 0.2 : 0.4 : 0.6 = 2 : 4 : 6 = 1 : 2 : 3	Multiply all of the numbers by 10 to make them integers. Then simplify by dividing by 2.	Another method is to find the multiplier that gives the simplest form in one step. What number would it be in this case?
b) $\frac{3}{4} : \frac{1}{8} : \frac{1}{4} = \frac{6}{8} : \frac{1}{8} : \frac{2}{8}$ = 6 : 1 : 2	Write the fractions with a common denominator, 8. Multiply all of the fractions by 8.	You can skip the first step and just multiply the fractions by 8. Try this.

Worked example 3

A soup contains 300 g of beans to every 1.5 kg of carrots. Express this ratio in its simplest form.

300 g : 1.5 kg = 300 g : 1.5 × 1000 g = 300 g : 1500 g = 1 : 5	Change both quantities to the same unit. 1 kg = 1000 g Simplify the ratio and write it without the units.	÷ 300 (300 : 1500) ÷ 300 1 : 5

Worked example 4

A paint called 'Spring pink' is made by mixing red paint and white paint in the ratio 2 : 3

A paint called 'Rose pink' is made by mixing red paint and white paint in the ratio 4 : 5

Which shade of pink is darker?

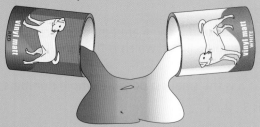

For Spring pink, red : white = 2 : 3 = 4 : 6 For Rose pink, red : white = 4 : 5	Rewrite the ratios so that both have the same number on one side. Compare 4 : 6 and 4 : 5	
Rose pink is darker.	For the same amount of red, Rose pink has less white than Spring pink.	For every 4 parts of red paint, Spring pink has 6 parts white paint and Rose pink has only 5 parts white paint:

Think about

With a calculator, how would you show which paint is darker using:

* the unitary method (making one side of each ratio equal 1)?
* the proportions of red in each pink paint?

Exercise 2   1–12

1. For each pattern, write the ratio of green : yellow : pink squares in its simplest form.

a)

b)

c)

d)

2 Express each ratio in its simplest form.

   a) 48 : 12 **b)** 25 : 60 **c)** 15 : 20 : 10 **d)** 18 : 15 : 21
   e) 24 : 12 : 36 **f)** 20 : 100 : 140 **g)** 24 : 8 : 30 **h)** 150 : 400 : 500

3 James gets a puzzle book. It contains 25 sudoku puzzles, 5 wordsearch puzzles and 10 crossword puzzles.

Write each ratio of puzzle types in its simplest form.

   a) sudoku : crossword **b)** crossword : wordsearch **c)** sudoku : crossword : wordsearch

4 Express each ratio in its simplest form.

   a) $\frac{2}{5} : \frac{1}{5} : \frac{4}{5}$ **b)** $\frac{1}{2} : 2\frac{1}{2} : 1$ **c)** 0.1 : 0.2 : 0.3 **d)** 0.4 : 0.6 : 0.8

   e) 1.2 : 7 : 3 **f)** 1.5 : 0.75 : 1 **g)** $\frac{2}{5} : \frac{1}{10} : \frac{3}{10}$ **h)** $\frac{1}{3} : \frac{5}{6} : \frac{2}{3}$

5 Aishwarya wants to write the ratio 0.25 : 0.4 : 0.8 in its simplest form. Her working is below.

$$0.25 : 0.4 : 0.8 = \frac{1}{4} : \frac{2}{5} : \frac{4}{5}$$

$$= \frac{5}{20} : \frac{8}{20} : \frac{16}{20}$$

$$= 5 : 8 : 16$$

Her answer is correct.

   a) Show a different method for simplifying 0.25 : 0.4 : 0.8

   b) Comment on which method is more efficient.

6 The ratio 1 hour : 60 minutes is a 1 : 1 ratio if you write both sides with the same unit. Write the missing number to complete each 1 : 1 ratio below.

   a) 1 cm : _____ mm **b)** 1 km : _____ m **c)** 1 m : _____ cm
   d) 1 kg : _____ g **e)** 1 litre : _____ ml **f)** 1 minute : _____ seconds

7 Simplify fully:

   a) 40 cm : 0.36 m **b)** 55 mm : 8 cm **c)** 800 g : 1.2 kg
   d) 6 weeks : 21 days **e)** 200 ml : 2 litres : 3 litres **f)** 3 minutes : 150 seconds : 2 minutes

8 Write the missing numbers to make equivalent ratios.

   a) 3 : 5 = 63 : ☐ **b)** 1 : 3 : 7 = ☐ : 18 : ☐ **c)** 2 : 5 : 9 = ☐ : ☐ : 72

   d) 3 : 4 : 5 = 0.3 : ☐ : ☐ **e)** 1 : 3 : 5 = ☐ : ☐ : 2.5 **f)** 2 : 3 : 4 = $\frac{1}{2} : \frac{☐}{4} : ☐$

9 The picture shows the ratio of red to blue marbles in five bags. Write them in order from the lowest to the highest proportion of red marbles.

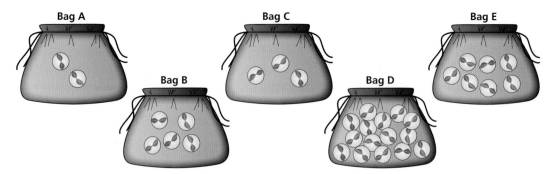

Bag A

Bag B

Bag C

Bag D

Bag E

10 Here are two fruit yogurts.

Fruit

Fruit : Yoghurt
2 : 5

Fruit

Fruit : Yoghurt
4 : 7

Find which of the yogurts contains the greater proportion of fruit. Show your working.

11 The aspect ratio of a photograph or a screen is the ratio of its length to its width, written in its simplest form.

a) Write down the aspect ratio of a square.

b) Ståle has a digital photograph which measures 1920 pixels by 1280 pixels. He wants to print it on paper that has aspect ratio 4 : 3. Explain whether this is possible without cutting the photo or the paper.

c) **Technology question** Use the internet to find out the most common aspect ratios for television screens and the aspect ratio used for 'widescreen' films at the cinema.

d) On some screens, a widescreen film does not fill the whole screen. Explain why.

12 A shop has two offers for two different products.

For each product, show which offer is the better value.

Product	Offer 1	Offer 2
Tea	$2.40 for a 500 g packet	$3.00 for a 500 g packet plus 20% extra
Coffee	$3.20 for a 200 g packet	$3.80 for a 200 g packet plus 25% extra

> **Tip**
>
> Either: write the ratio of cents to grams for each offer, and then make the numbers the same on one side of the ratios; or calculate the cost per gram for each offer.

13 A shop has two offers for two different products.

For each product, show which offer is the better value.

Product	Offer 1	Offer 2
Rice	$5.50 for a 5000 g bag	$7.50 for a 7 kg bag
Orange juice	$0.84 for 750 ml	$2.85 for 2.5 litres

Discuss

What are the possible methods for using a calculator to answer question 13?

Thinking and working mathematically activity

A model is made of blue, green, red and yellow toy bricks.
The numbers of bricks are in the following ratios:

blue : green = 3 : 2

green : red = 3 : 2

red : yellow = 3 : 2

Find the minimum total number of bricks in the model.

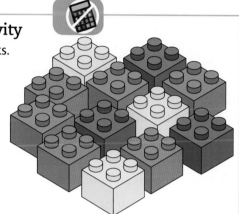

22.3 Dividing a quantity into a ratio

Worked example 5

Share 580 g in the ratio 2 : 3 : 5

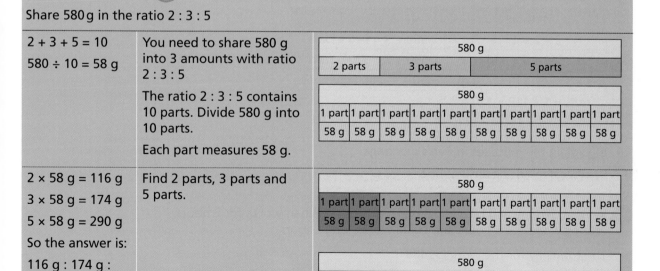

2 + 3 + 5 = 10 580 ÷ 10 = 58 g	You need to share 580 g into 3 amounts with ratio 2 : 3 : 5 The ratio 2 : 3 : 5 contains 10 parts. Divide 580 g into 10 parts. Each part measures 58 g.
2 × 58 g = 116 g 3 × 58 g = 174 g 5 × 58 g = 290 g So the answer is: 116 g : 174 g : 290 g	Find 2 parts, 3 parts and 5 parts.

Exercise 3

1 Divide $480 into these ratios.

   a) 3 : 2   **b)** 5 : 7   **c)** 1 : 2 : 3   **d)** 3 : 5 : 8   **e)** 6 : 7 : 7   **f)** 2 : 9 : 13

2 June has 30 days. The ratio of snowy days : rainy days : dry days in is 3 : 2 : 5

   a) Find the number of dry days.

   b) How many more days were snowy than rainy?

3 Ebony spends an hour exercising. She does walking, running and stretching in the ratio 4 : 5 : 3

   Find the number of minutes she spends on each type of exercise.

4 The ratio of angles in a triangle is 7 : 3 : 5. Show whether or not the triangle is right-angled.

5 A cereal contains a mixture of oats, nuts and wheat in the ratio of 3 : 2 : 1.
   A box contains a total of 660 g. To find the mass of oats, nuts and wheat in the box,
   Ben uses two different methods. His answers are both wrong.

Method 1	Method 2
3 + 2 + 1 = 6	660 ÷ 3 = 220 g oats
660 ÷ 6 = 110	660 ÷ 2 = 330 g nuts
oats = 2 × 110 = 220 g	660 ÷ 1 = 660 g wheat
nuts = 1 × 110 = 110 g	
wheat = 3 × 110 = 330 g	

   a) Explain the mistakes in Ben's workings.

   b) What should the answer be?

6 Reyhan, Jeremy and Leida do a piece of work. Reyhan works for 7 hours,
   Jeremy works for 5 hours and Leida works for 11 hours.

   The total pay for the work is $483.

   Find the amount of money that each person should get.

Thinking and working mathematically activity

A bag contains blue, green and red marbles. The ratio of blue to green to red marbles is 2 : 3 : x.
There are 60 marbles in the bag.
Find all of the possible values of x. How do you know you have found them all?
Find the largest and smallest possible numbers of red marbles in the bag.

Consolidation exercise   1–8

1 Match each description to the correct ratio and proportion.

Complete the missing boxes to produce six matching trios.

Description	Ratio A : B	Proportion
A and B share 40. A receives 15.		A : $\frac{1}{4}$ B : $\frac{3}{4}$
A receives 5 and B receives 15.	3 : 1	
A and B share 32. B receives 12.	2 : 1	A : $\frac{5}{8}$ B : $\frac{3}{8}$
A receives 14 and B receives 7.	1 : 3	A : $\frac{2}{5}$ B : $\frac{3}{5}$
A and B share 65. A receives 26.	2 : 3	A : $\frac{3}{8}$ B : $\frac{5}{8}$
	5 : 3	A : $\frac{3}{4}$ B : $\frac{1}{4}$

2 Graham is dividing some money between two people in the ratio 2 : 3

He says, 'The first person will get $\frac{2}{3}$ of the money.'

Is Graham correct? Explain your answer.

3 Tom and Mel share some money between them. Tom receives $\frac{2}{7}$ of the money.

Mel receives $125. How much money does Tom receive?

4 Sesilia makes bracelets. She has 125 red beads, 60 blue beads and 315 yellow beads.

Write the ratio of red to blue to yellow beads in its simplest form.

5 Write each ratio in its simplest form.

a) \$2 : 150 cents = _____ : _____   **b)** 1000 g : 3 kg = _____ : _____

c) 3 hours : 180 minutes = _____ : _____   **d)** 32 cm : 0.8 m = _____ : _____

6 Find the missing integer in each box.

a) 27 : 45 : ☐ = 3 : ☐ : 2   **b)** 0.9 : 1.5 : 3 = ☐ : 5 : ☐

> **Tip**
>
> In part **b)**, think about how to get from 1.5 to 5 in two steps.

7 Greta, Xavier and Tim grow yams in a garden.
They share 48 yams in the ratio 3 : 1 : 4

Find the number of yams each person gets.

8 In a large hotel's kitchen, the ratio of chefs to assistants to cleaners is 1 : 5 : 3
If there are 5 chefs, find the numbers of assistants and cleaners.

9 The table shows the number of visitors to a museum on one weekend.

	Adults	Children
Saturday	720	580
Sunday	850	600

Tom thinks that the larger proportion of children visiting is on Sunday, as more children visited on Sunday than on Saturday. Is he correct? Explain your answer.

End of chapter reflection

You should know that...	You should be able to...	Such as...
A ratio compares parts to each other. A proportion compares one part with the whole.	Find proportions from a ratio and find a ratio from a proportion.	In a school, the ratio of teachers to students is 2 : 15. Write the proportion of teachers and the proportion of students. Jo and Athar share a pizza. Jo eats $\frac{2}{5}$ of the pizza and Athar eats the rest. Find the proportion that Athar eats, and the ratio of the amounts Jo and Athar eat.
A ratio can have more than two parts.	Simplify ratios with more than two parts, including ratios with parts that are fractions or decimals.	Write each ratio in its simplest form. **a)** 15 : 27 : 12 **b)** 0.8 : 0.2 : 6 **c)** $\frac{3}{2} : \frac{1}{4} : \frac{1}{2}$

To simplify a ratio whose parts have different units, first convert the parts to the same unit.	Simplify ratios with parts that have different units.	Write each ratio in its simplest form. **a)** 6 hours : 2 days **b)** 0.5 kg : 750 g : 1200 g
To compare the proportions of a quantity in two different ratios, you can: • multiply the ratios so that one side is the same in both • divide the ratios so that one side equals 1 in both • use the numbers in the ratios to directly calculate the proportions of the quantity.	Show which of two ratios shows a greater proportion of one of the quantities.	Two pink paints are made from mixing red and white paint in the following ratios: paint A 9 : 13 paint B 27 : 37 Show which paint is darker.
A quantity can be divided into a ratio with more than two parts.	Divide a quantity into a ratio with more than two parts.	Share 500 ml in the ratio. 8 : 3 : 14

23 Relationships and graphs

You will learn how to:

- Use knowledge of coordinate pairs to construct tables of values and plot the graphs of linear functions, where y is given explicitly in terms of x ($y = mx + c$).
- Recognise that equations of the form $y = mx + c$ correspond to straight-line graphs, where m is the gradient and c is the y-intercept (integer values of m).
- Understand that a situation can be represented either in words or as a linear function in two variables (of the form $y = mx + c$) and move between the two representations.
- Read and interpret graphs with more than one component. Explain why they have a specific shape and the significance of intersections of the graphs.

Starting point

Do you remember...

- how to substitute into a formula?

  For example, substitute $x = 4$ into the formula $y = x - 2$

- how to generate coordinate pairs that satisfy a simple linear equation and draw the graph?

  For example, copy and complete this table to give the coordinate pairs for $y = x - 3$ for $x = 0, 1, 2, 3$:

x	0	1	2	3
y				

- how to give the equation of a vertical or horizontal line?

  For example, give the equation of the vertical line that passes through the point (4, 5).

This will also be helpful when...

- you find the approximate solutions of a simple pair of simultaneous linear equations by finding the point of intersection of their graphs.

23.0 Getting started

Coordinate Battleships

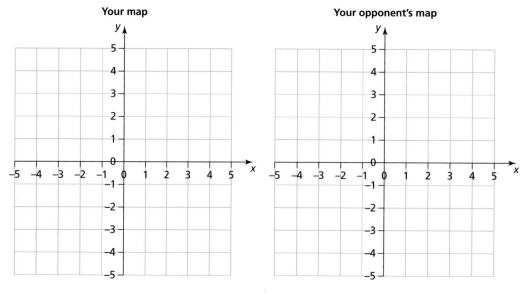

On your map, you will use the grid lines to draw 7 boats.

You need two boats of 1 square in length, 2 squares in length and 3 squares in length and one boat of 4 squares in length (as shown in the diagram).

Your boats must be either horizontal or vertical (not diagonal) and must be placed on the grid lines.

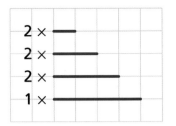

Now decide who is starting.

Try to guess where your opponent's boats are by guessing a coordinate.

If the coordinate that you choose hits one of your opponent's boats, mark this coordinate with a cross. If it does not hit a boat, mark this with a circle. Do this on your copy of your opponent's map.

Your opponent now has their turn to guess where your boats are hidden.

To sink your opponent's boats, give the equation of the line that the boat is on. Keep playing until one player has sunk all of their opponent's boats.

23.1 Plotting graphs of linear functions

Key terms

A **linear function** is any function that graphs to a straight line.

All linear functions can be written in the form $y = mx + c$, where m and c are constants.

Worked example 1

Here is the equation of a line: $y = 2x - 1$

a) Complete a table of values for values of x between −2 and 3.

b) Plot the graph of $y = 2x - 1$ for values of x between −2 and 3.

a)

x	−2	−1	0	1	2	3
y	−5	−3	−1	1	3	5

You need to use the values of x to calculate the corresponding values of y.

For example, when $x = -2$, $y = 2 \times (-2) - 1$ $= -5$.

This tells you that the point (−2, −5) is a coordinate on the line.

Similarly, other coordinates of points on the line are (−1, −3), (0, −1), (1, 1), (2, 3) and (3, 5).

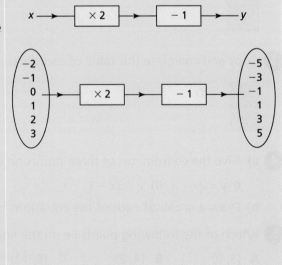

b)

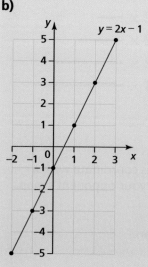

$y = 2x - 1$

Plot each coordinate that you have found on your axes.

Join up the points. If they do not form a straight line then check your calculations. If they do form a straight line, then label it with the equation.

Think about

How do you know what values the y-axis must be drawn between?

Discuss

What is the minimum number of coordinates you need in order to plot a line?

Exercise 1

1 Copy and complete the missing coordinates of points on the line $y = 4x + 1$

a) (2, ...) b) (5, ...) c) (0, ...) d) (1, ...) e) (–2, ...)

2 Copy and complete this table of coordinates for the equation $y = 3x + 7$

x	0	1	2	3	4	5
y						

3 Copy and complete this table of coordinates for the equation $y = 2x - 4$

x	0	1	2	3	4	5
y						

4 a) Give the coordinates of three points on each of these lines. Use values of x from 0 to 6.

i) $y = 2x + 5$ ii) $y = 3x - 1$

b) Draw a graph of each of the equations in part **a**. Use values of x from 0 to 6.

5 Which of the following points lie on the line $y = 2x - 6$?

A (3, 0) B (4, 2) C (8, 12) D (15, 26) E (20, 34)

6 Draw a table of values for each of the equations below and then draw the graph for each equation on a separate set of axes.

x	−3	−2	−1	0	1	2	3
y							

a) $y = 3x + 1$

b) $y = 4x − 3$

c) $y = 2x + 3$

7 By drawing a table of values, draw each of these graphs.

a) $y = 2x − 5$ b) $y = 3x − 4$ c) $y = 4x + 1$

8 For each part, draw a table of values for values of x between −3 and 3 and then plot the graphs on separate axes.

a) $y = 3 − x$ b) $y = 0.5x$ c) $y = 1 − 2x$ d) $y = 3 − 4x$

9 Jessie and Ben use two different ways to draw the line $y = 6 − 3x$.

Jessie copies and completes the table of coordinates for the line. She joins the points.

x	−3	−2	−1	0	1	2	3
y	15	12	9	6	3	0	−3

Ben finds the points where the line crosses the x- and y-axis. He joins the two points.

$y = 6 − 3x$ $y = 6 − 3x$

$x = 0$ $y = 0$

$y = 6 − 3 × 0 = 6$ $6 − 3x = 0,$ $3x = 6$

$y = 6$ $x = 2$

$(0, 6)$ $(2, 0)$

Jessie's graph

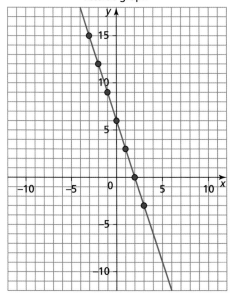

Ben's graph

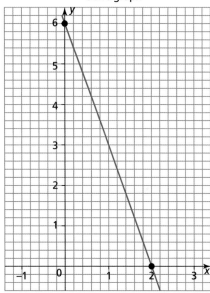

a) Compare the methods that Jessie and Ben have used. What are the advantages and disadvantages of each method?

b) Use both methods to draw the graph of $y = 4 − 2x$.

10 Use the method that Ben used in question 9 to draw the graphs of:

a) $y = 3x + 6$ b) $y = 2x - 4$ c) $y = 2 - 4x$ d) $y = -10 + 5x$

Thinking and working mathematically activity

Is each statement below true or false? Explain your reasoning. Draw lines to support your answers.

- To draw a line, I need to complete a table of values for x from -3 to 3.
- To draw a straight line, I need to know the coordinates of at least one point.
- Two different lines can cross the x- and y-axis at the same points.
- I cannot draw a line if I do not know where the line crosses the two axes.

23.2 Equation of a line

Key terms

The **gradient** of a line is a measure of how steep it is.

If the equation of the line has the form $y = mx + c$, m represents the gradient.

A formula for the gradient of a line is:

$$\text{gradient} = \frac{\text{increase in } y\text{-coordinates}}{\text{increase in } x\text{-coordinates}}$$

The **intercept** is where the line crosses the x- and y-axis.

The x-intercept is where the line crosses the x-axis.

The y-intercept is where the line crosses the y-axis.

Worked example 2

Find the gradient of each line.

a)

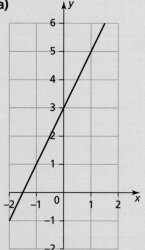

b)

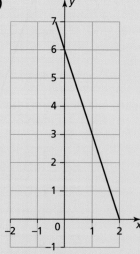

a) The gradient is $\frac{4}{2} = 2$

This graph has a positive gradient as the line slopes upwards (looking from left to right).

Identify two points that the line passes through and find the change in the y-coordinates and the change in the x-coordinates.

Find the gradient using:

$$\frac{\text{increase in } y\text{-coordinates}}{\text{increase in } x\text{-coordinates}}$$

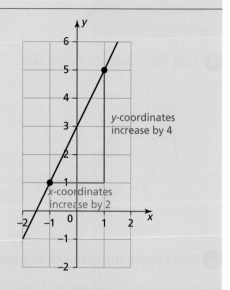

y-coordinates increase by 4

x-coordinates increase by 2

b) The gradient is $\frac{-3}{1} = -3$

This graph has a negative gradient as the line slopes downwards.

This graph goes <u>down</u> by 3 when the x-values increase by 1.

This is the same as saying that the increase in the y-coordinates is −3.

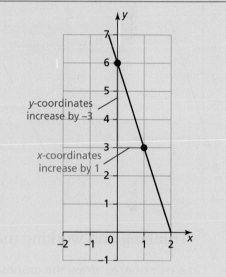

y-coordinates increase by −3

x-coordinates increase by 1

Tip

The graph in part **b)** has:

* y-intercept of 6 (as this is the value where the line cuts through the y-axis)
* x-intercept of 2 (as this is the value where the line cuts through the x-axis)

1 Find the gradient of each line.

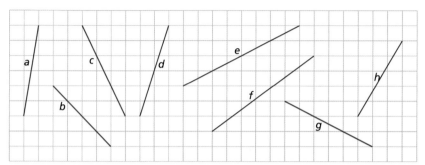

2 Write down the gradient and the *y*-intercept for each line.

a)
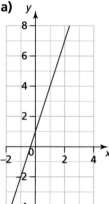

b)
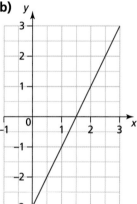

c)
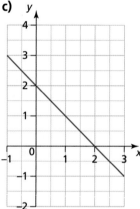

d)
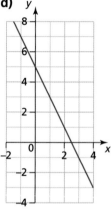

▼ **Thinking and working mathematically activity**

• On a pair of axes, draw the graphs of $y = 2x - 1$, $y = 2x + 2$ and $y = 2x + 3$

What is the same and what is different about the lines?

Find the *y*-intercept of each line.

Predict what the *y*-intercept of the line $y = 2x + 5$ will be.

• On a pair of axes, draw the graphs of $y = 3x + 1$, $y = x + 1$ and $y = 1 - 2x$

What is the same and what is different about these lines?

Find the gradient of each line.

Predict what the gradient of the line $y = 4x + 1$ will be.

• Write down some conclusions. How can you tell what the gradient and *y*-intercept of a graph will be from its equation?

---

3 For each equation, write down the gradient and the *y*-intercept.

a) $y = 6x + 3$

b) $y = 3x - 8$

c) $y = 5x + 1$

d) $y = x - 4$

e) $y = -3x + 4$

f) $y = 3 - x$

g) $y = 5 + 6x$

h) $y = 4$

---

4 Write down the equations from the box that are **not** equations of straight line graphs.

> $y = 3x$ $y = -x$ $y = 5$
>
> $y = 2x^2$ $y = 0$ $y = 6 - 4x$
>
> $x = 1$ $x = 7.5$ $y = \dfrac{6}{x}$

5 a) Match each equation to the correct graph.

A $y = x$ **B** $y = -x$ **C** $y = 3 - x$ **D** $y = 3$ **E** $y = 3 + x$

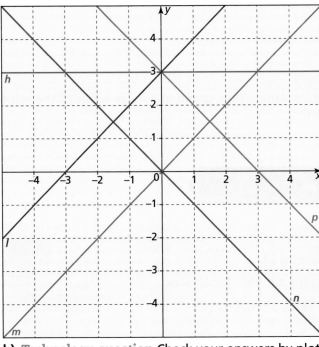

b) Technology question Check your answers by plotting the five lines using graph drawing software.

6 Some straight-line graphs have been sorted in a table.
Copy and complete possible headings for the table.

Lines with	Lines with ....
$y = 3 - 2x$	$y = 4x$
$y = 3$	$y = 4x + 1$
$y = -4x + 3$	$y = 3 + 4x$
$y = 0.5x + 3$	$y = -10 + 4x$

7 Write the following equations in the appropriate place on the Venn diagram.

A $y = 3x + 1$ **B** $y = 7 - 2x$ **C** $y = 1 - x$ **D** $y = x + 3$ **E** $y = 2x$

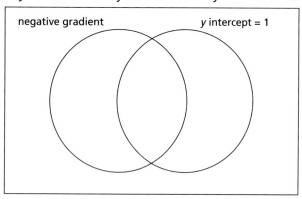

negative gradient

y intercept = 1

> **Think about**
>
> Sabrina thinks that the graph of $y = 2x - 3$ is the same as the graph of $y = -2x + 3$ as all the values have been multiplied by -1. Is Sabrina correct?

23.3 Real-life graphs and functions

Worked example 3

The charge for using a car park is $3 per hour, plus $1.

a) Write a function for the cost, $C, for using the car park for t hours.

b) Complete the table of values for the cost of the car park.

Time, t (hours)	0	2	4	6
Cost, C ($)	1	7		

c) Plot the graph for the cost of the car park.

d) Amir pays $25 for the car park. How many hours did he leave his car in the car park?

a) The cost, $C, for using the car park for t hours can be found using $C = 1 + 3t$	The car park is $3 per hour, so for t hours this is $3t$ There is also a fixed cost of $1 The total cost is $(3t + 1)$	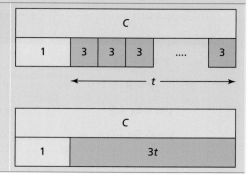

b)

Time, t (hours)	0	2	4	6
Cost, C ($)	1	7	13	19

When t = 4, the cost is:

3 × 4 + 1 = 12 + 1 = $13

When t = 6, the cost is:

3 × 6 + 1 = 18 + 1 = $19

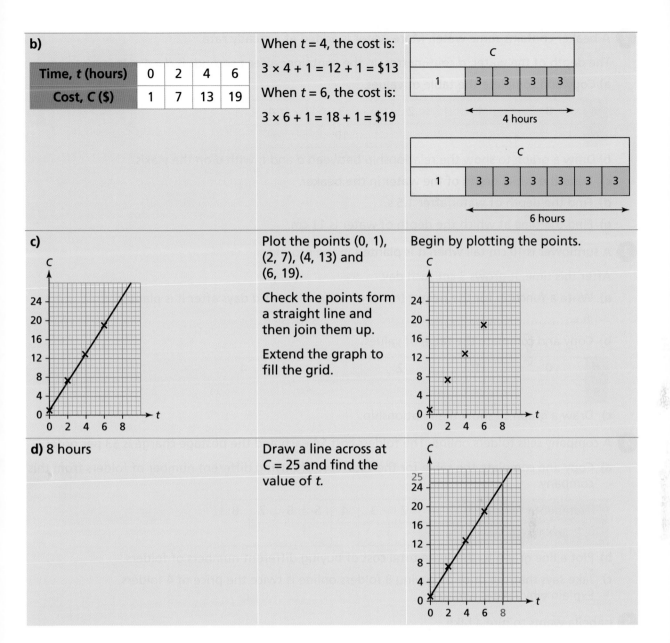

c)

Plot the points (0, 1), (2, 7), (4, 13) and (6, 19).

Check the points form a straight line and then join them up.

Extend the graph to fill the grid.

Begin by plotting the points.

d) 8 hours

Draw a line across at C = 25 and find the value of t.

Exercise 3

1 The temperature of a liquid starts at 100 °C.

The temperature decreases by 5 °C each minute.

a) Copy and complete the table for the temperature of the liquid over time.

Time (minutes)	0	1	2	3	4	5	6	7	8
Temperature (°C)	100								

b) Draw a graph to show the results.

c) Use your graph to find after how many minutes the temperature had fallen to 77.5 °C.

2 A beaker contains some water. More water is added at a steady rate.

The depth of the water, d centimetres, in the container after t seconds is $d = 4 + 2t$

a) Copy and complete the table of values.

t	0	1	2	3	4
d				10	

b) Draw a graph to show the relationship between d and t, with d on the y-axis.

c) State the initial depth of the water in the beaker.

d) Find the depth of water after 1.5 s.

e) Find the time at which the depth of water is 11 cm.

3 A sunflower is 10 cm tall when it is planted in the ground.

After this time, it grows 3 cm each day.

a) Write a function for the height (h cm) of the sunflower d days after it is planted.

   $h = $ ....

b) Copy and complete the table of values.

d	0	1	2	3	4
h					

c) Draw a graph to show the relationship.

4 A company sells folders online. The folders cost $4 each and the postage charge is $3 per order.

a) Copy and complete the table for the total cost of buying different number of folders from this company.

Number of folders, x	1	2	3	4	5	6	7	8
Total cost, $y								

b) Plot a line graph to show the total cost of buying different numbers of folders.

c) Jake says that the price of buying 8 folders online is twice the price of 4 folders. Explain why Jake is wrong.

5 Isabella wants to hire a bike.

She can hire a bike from two companies.

The table shows some information about how much each company charges.

	Description in words	Function (y is the cost for hiring a bike for x hours)
Company A	The charge is $2 per hour, plus $8	
Company B		$y = 3x + 5$

a) Copy and complete the table.

Isabella decides to use Company A.

b) Draw a graph to show the cost of hiring a bike from Company A for up to 5 hours.

c) Company A charges Isabella $15 for hiring a bike. Use your graph to find the number of hours she hired the bike for.

6 A car has 56 litres of fuel in its fuel tank. It uses 7 litres of fuel per hour of driving.

    a) Write a formula for the volume, *V* litres, of fuel in the car's tank after *t* hours of driving.

    b) Draw a graph to show the relationship between *V* and *t* for $0 \leq t \leq 7$.

    c) Why is the graph not reliable for values of *t* greater than 8?

7 Tamara sells pizzas. Customers choose how many toppings they would like on their pizza.

    She uses the formula $y = 2x + 3$ to find the cost ($*y*) of a pizza with *x* toppings.

    Tamara draws a graph to show the relationship between *y* and *x*.

    a) Write down the *y*-intercept of the graph.

    b) How much is the cost of one pizza if a customer has no toppings?

    c) Write down the gradient of the line.

    d) What is the cost of each topping?

8 Ben rents a room to run drawing classes for children. He uses the equation $P = 8c - 25$ to work out his profit, $*P*, when *c* children attend the class.

    a) Draw the graph of the line $P = 8c - 25$ for $0 \leq c \leq 6$

    b) Write down the value where the line intercepts the *P*-axis.
State what it represents in this situation.

    c) Find the gradient of the line and state what it represents in this situation.

9 A company uses the equation $y = 15x + 100$ to work out the monthly pay ($*y*) for their staff, where *x* is the number of hours of work.

    a) Draw the graph of this equation for $0 \leq x \leq 100$.

    The payment included a basic payment and then an amount for every hour worked.

    b) Write down the basic payment.

    c) Write down the hourly rate of pay.

10 The graph shows the total amount ($*F*) in an investment fund after *t* years.

    a) Write down the initial value of the investment.

    b) Write down the amount that the investment increases by each year.

    c) Layla says that the equation of the line is $t = 5000 + 3000F$. Explain why she is incorrect.

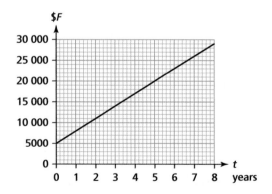

11 Two mobile phone companies offer phone contracts which include unlimited text and calls but not internet use.

Company P charges a fixed monthly fee of $5, plus $20 for every gigabyte of data downloaded.

Company Q charges a fixed monthly fee of $10, plus $10 for every gigabyte of data downloaded.

a) Write a formula for the total monthly cost, $c, when the amount of data downloaded is d gigabytes:

i) with company P   ii) with company Q.

b) On the same axes, draw graphs to show the monthly cost with company P and with company Q, for up to 4 gigabytes downloaded.

c) Use your graph to find the number of gigabytes for which both companies charge the same amount.

d) Find the difference between the monthly costs of the two companies if you download 3 gigabytes of data.

 Thinking and working mathematically activity

Tariq investigates the cost of hiring a car.

He discovers that the cost (c dollars) of hiring a car for a day depends on the distance driven (d kilometres).

The table shows the formulae six companies use to calculate the cost.

Company	Formula
A	$c = 90$
B	$c = 2d + 50$
C	$c = 30 + 3d$
D	$c = 2d$
E	$c = 30 + d$
F	$c = 4d - 10$

Tariq draws a graph for each company showing the cost plotted against the distance driven.

Decide if each of the following statements is true or false. Give a reason for each answer.

• Company A has the steepest graph.
• Only one of the companies has a graph that passes through the origin.
• Company E is the cheapest company for a distance of 35 km.
• One of the companies has a graph that has a negative gradient.
• Two of the companies have the same flat rate fee.
• Two of the companies charge the same when a distance of 10 km is driven.
• Company C is sometimes the cheapest company.

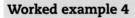

Key terms

....

Constant speed is when an object travels at a steady pace without speeding up or slowing down.

A **real-life graph** compares changes over time. For example, a travel graph shows how distance changes over time.

Worked example 4

Susan and Owen run to the end of the road, rest and then race back to the start.

Owen twisted his ankle on the way back. The graph shows the full race for both Susan and Owen.

a) Who completed the race in the shortest time?

b) How long did Susan rest for?

c) After how many seconds did Owen twist his ankle?

d) What happened 68 seconds after the start of the race?

e) What was the total distance of the race?

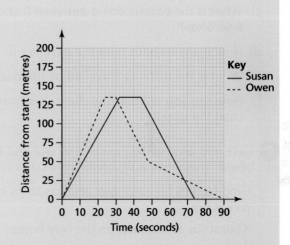

a) Susan completed the race first. Since they both start at 0 metres, the race will end when they get back to 0 metres.	It is clear that Susan's line arrives back at 0 metres before Owen's. 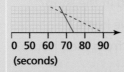
b) 12 seconds Read the scale carefully. Each smaller square represents 2 seconds. 6 small squares, so 6 × 2 = 12	The horizontal line represents the stationary part of race when they rested. 12 seconds
c) 48 seconds after the start of the race.	Owen's speed significantly decreased at 48 seconds. This is shown by the graph becoming less steep.
d) Susan overtook Owen.	The lines cross when they are exactly the same distance away from the start/finish.
e) 270 metres 2 × 135 = 270	The distance to the end of road is 135 m and so the full distance of the race is 2 × 135 since they need to run back to the start.

1 A cheetah is beneath a tree when it spots a gazelle.

The cheetah then hunts the gazelle. The graph shows the chase.

a) How far away from the gazelle is the cheetah when the gazelle starts to run?

b) What happens at 11.5 seconds?

c) What is the gazelle doing between 0 and 5 seconds?

d) How many seconds does the gazelle run for?

e) What is the distance travelled by the gazelle?

f) How can you tell that the cheetah is faster than the gazelle?

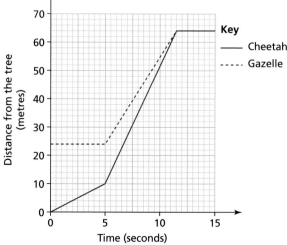

2 Kate and Gino go to the same school and live in the same house. They both walk home from school one day.

Gino forgets his lunch box and has to go back to school to pick it up.

Kate stops at the shop on the way home.

Both of their journeys are shown on the travel graph.

a) Complete the key to show which line represents each person.

b) How long does Kate spend in the shop?

c) What time does Kate arrive home?

d) How far away is the shop from their house?

e) What happens at 14:46?

f) Who walks at the faster pace? Give a reason for your answer.

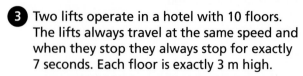

3 Two lifts operate in a hotel with 10 floors. The lifts always travel at the same speed and when they stop they always stop for exactly 7 seconds. Each floor is exactly 3 m high.

The first lift starts at ground level. It goes up to the 2nd floor, then goes up to the fifth floor, then goes down to the fourth flour and then goes back down to the ground floor.

This journey is shown on the travel graph.

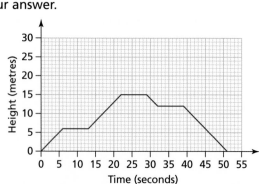

A second lift starts on the tenth floor at same time as the first lift starts at ground level. It goes down to the second floor, then goes up to the fourth floor and then moves up to the sixth floor.

a) Copy the journey of the first lift on squared paper, and add on the journey for the second lift.

b) For how many seconds are both of the lifts on the same floor at the same time?

4 The travel graph shows journeys made by two lorries.

Lorry 1 travels from a factory to a shop.

Lorry 2 travels from the shop back to the factory.

Both lorries use the same roads.

a) What is the distance between the factory and the shop?

b) At what time does Lorry 2 arrive at the factory?

c) The driver of Lorry 2 stopped for a break at 12:00. Write down the length of this break.

d) Find the distance between the two lorries at 13:00

e) Estimate the time when the two lorries passed each other.

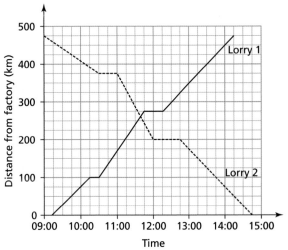

5 Four bicycle hire companies each have a different way of charging. Each graph shows a relationship between the cost, c dollars, of bicycle hire, and the time, t days, for which a bicycle is hired.

a) Match each company with the correct graph.

Company P charges no fixed fee.
There is a cost per day to hire a bicycle.

Company Q charges a fixed fee which includes the first three days of bicycle hire. After three days, there is an added cost per day.

Company R charges a cost per day and no fixed fee. After the third day, the cost per day changes to a lower value.

Company S charges a fixed fee, plus a cost per day.

Graph 1

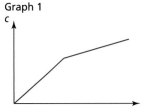

Graph 2

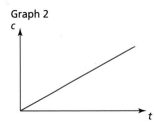

Graph 3

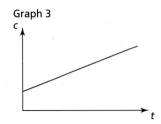

Graph 4

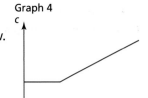

b) State which of the graphs could match each of the functions below.

i) $c = 20t$ **ii)** $c = 15 + 15t$

6 The graphs show the journey of two trains. Both trains leave from the same station.

Look at the statements below describing the journeys. Some of the statements are not correct.

Copy each statement, correcting any facts that are not correct.

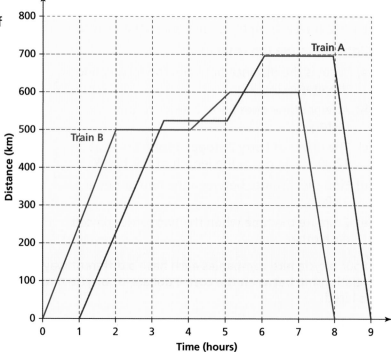

Train A travelled a total distance of 700 km during this time period.

Train B left 1 hour after Train A.

Train A arrived back at its starting point 8 hours after it left.

Train B travelled faster than Train A on the journey back to the starting station.

Both trains stopped the same amount of time throughout the journey.

 Thinking and working mathematically activity

- Draw a travel graph to help solve this problem.

  A train travels from Station A to Station B, a distance of 60 km.
  The train leaves Station A at 16:00
  It stops at a signal at 16:40 when it is 36 km from Station A.
  After 10 minutes it starts moving again and it arrives at Station B at 17:20

  A second train leaves Station B at 16:30 and travels to Station A without stopping.
  This second train arrives at Station A at 17:35

  Find how far the trains are from Station A when they pass.

- Make up your own similar problem.

  Give your problem to a partner to solve.

Some travel graphs are made from straight lines and some are curves.

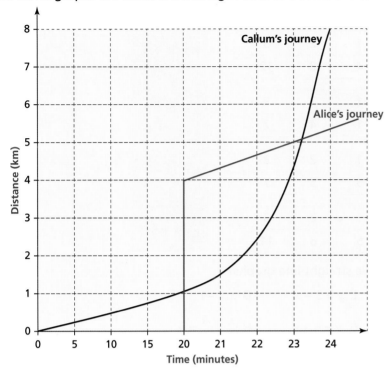

- Why is the graph showing Alice's journey unrealistic?
- Callum's graph is an increasing curve. What does this tell you about his journey?

Consolidation exercise

1 Copy and complete this table.

Equation of line	Gradient of line	y-intercept
$y = 7x - 4$	……..	−4
$y = 3x + $ …….	……..	5
$y = 9x + 6$	……..	………
$y = -4x + 10$	……..	………
………	6	−5

2 Three of these lines pass through the same point.

Which of these lines does not pass through the point?

A: $x = 4$ B: $y = 4$ C: $y = 4x$ D: $y = 2x - 4$

3 Match each linear equation with the correct table of values.

A

x	−2	−1	0	1	2	3
y	3	4	5	6	7	8

B

x	−2	−1	0	1	2	3
y	−8	−4	0	4	8	12

C

x	−2	−1	0	1	2	3
y	−8	−5	−2	1	4	7

D

x	−2	−1	0	1	2	3
y	−3	−1	1	3	5	7

E

x	−2	−1	0	1	2	3
y	−4	−1	2	5	8	11

$y = 4x$

$y = 2x + 1$

$y = 3x + 2$

$y = 3x − 2$

$y = x + 5$

4 Which of these equations will give straight line graphs?

A $y = 3x − 4$ **B** $y = 1$ **C** $y = 3 − x$ **D** $x = −1$

E $y = x^3$ **F** $y = 0.5x$ **G** $y = 6 − x$ **H** $\dfrac{x}{3} − \dfrac{4}{3} = y$

5 A ferry set off from Dover to Calais at 00:39. A few minutes later, a second ferry set off from Calais to Dover.

As the first ferry approached Calais it needed to stop and wait for clearance before docking.

The graph shows the journeys of both ferries.

a) What time did the second ferry set off from Calais to Dover?

b) What is the distance between Dover and Calais?

c) Were the ships closer to Dover or to Calais when they passed each other?

d) What was the journey time of the second ferry?

e) How long did the first ferry have to stop for before docking?

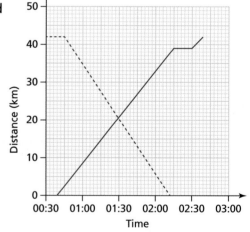

6 A small aeroplane has 5500 litres of fuel in its tank. It uses 1000 litres of fuel per hour of flying.

a) Write a formula for the volume, V litres, of fuel in the aeroplane's tank after t hours of flying.

b) Draw a graph to show the relationship between V and t, for $0 \le t \le 4$

c) Use the graph to find the volume of fuel in the tank after 3.5 hours.

7 Nathan and Julie are both running a metric marathon which is about 26 km in length.

Nathan runs at a constant speed and completes the marathon in 4 hours and 15 minutes.

Julie has two rests which total 24 minutes, before completing the marathon in 4 hours and 42 minutes.

a) Draw a distance time graph for Nathan's marathon.

b) Draw a possible distance time graph for Julie's marathon.

End of chapter reflection

You should know that...	You should be able to...	Such as...
The points that lie on a line can be described using an equation.	Plot a graph of a given equation.	Plot the graph of $y = 2x - 1$ for integer values of x between −3 and 3.
The gradient measures the steepness of a line.	Find the gradient of a straight line.	Find the gradient of this line.
Straight-line graphs have an equation in the form $y = mx + c$.	Say whether a graph is a straight-line graph from the equation of the line.	Which of the following equations are not straight lines? Explain your reasoning. **a)** $y = 3x + 1$ **b)** $y = x^2$ **c)** $y = 4 - 2x$ **d)** $y = 4$ **c)** $x = -2$
	m represents the gradient of the line and c is the value of the y-intercept.	Write down the gradient and the y-intercept of the line $y = 4x - 1$

A graph can be used to show a real-life relationship between two variables.	Draw a graph, given a description of a linear relationship.	A candle with a height of 20 cm burns down at a rate of 1.5 cm per hour. Draw a graph to show the relationship between the time, t minutes, and the candle's height, h centimetres.
	State the meaning of the y-intercept and the gradient of a graph of a real-life function.	Say what the y-intercept and the gradient of the candle graph (above) represent.

Thinking statistically

You will learn how to:

- Record, organise and represent categorical, discrete and continuous data.
  Choose and explain which representation to use in a given situation:
  - Venn and Carroll diagrams
  - tally charts, frequency tables and two-way tables
  - dual and compound bar charts
  - pie charts
  - frequency diagrams for continuous data
  - line graphs and time series graphs
  - scatter graphs
  - stem-and-leaf diagrams
  - infographics.
- Use knowledge of mode, median, mean and range to compare two distributions, considering the interrelationship between centrality and spread.
- Interpret data, identifying patterns, trends and relationships, within and between data sets, to answer statistical questions. Discuss conclusions, considering the sources of variation, including sampling, and check predictions.

Starting point

Do you remember...

- how to draw and interpret different statistical graphs?

  For example, what percentage of people represented in this pie chart do not eat lunch?

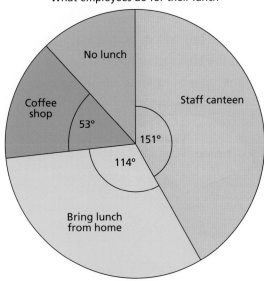

What employees do for their lunch

- how to calculate the mean, median, mode and range?

  For example, find the median, mean and range of these times (given in seconds).

  | 11.4 | 10.9 | 11.2 | 10.8 | 11.0 |

  | 11.5 | 10.9 | 11.3 | 11.7 | 11.5 |

This will also be helpful when...

- you work on your own statistics project
- you compare data presented in other types of statistical graphs.

24.0 Getting started

This is an activity for two players to test your reaction time.

Hold a ruler vertically above your opponent's hand, with their arm resting on the desk.

Drop the ruler without saying anything. Your opponent must then catch the ruler as quickly as possible. Record the length at which they caught it.

Repeat this activity a number of times. The winner is the one with the best average.

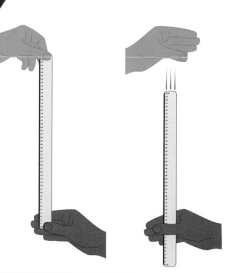

- Decide how many times you will repeat the game.
- Decide whether you will use the mean, median or mode.
- Display the results in a suitable way.
- Did the winner have the most consistent results?

Think about

How do you think reaction times might change with age? See if you can find out.

24.1 Comparing and using graphs

Worked example 1

A company that runs holiday cruises has two ships, the Starlight and the Suncatcher.

The Starlight has 120 cabins and the Suncatcher has 600 cabins.

The pie charts show the different types of cabin available on each of the ships.

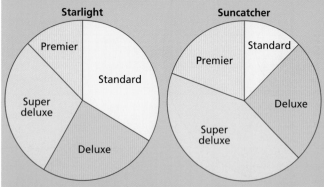

a) Write down three comparisons based on the pie charts.

b) Blessy says, 'There are more standard cabins on the Starlight than on the Suncatcher.'
   Do the pie charts support what Blessy says?

a) Standard is the most common type of cabin on the Starlight, but Super deluxe is the most common type of cabin on the Suncatcher.

The Deluxe cabin was the same proportion of the cabin types on both the Starlight and the Suncatcher.

Premier is the least common type of cabin on the Starlight, but Standard is the least common type of cabin on the Suncatcher.

The largest sector of each pie chart shows the most frequent type of cabin for that ship.

The pie charts show the proportion of the total number of cabins that are of each type. The sector for Deluxe is the same fraction of the pie chart for both Starlight and Suncatcher.

The smallest sector of each pie chart shows the least frequent type of cabin for that ship.

b) Blessy has not taken into account the different numbers of cabins represented by each of the pie charts. Standard cabins represent a greater proportion of the cabins on Starlight than on Suncatcher, but there are more cabins on Suncatcher.

Around $\frac{1}{3}$ of the cabins on Starlight are standard cabins. $\frac{1}{3}$ of 120 = 40

Around $\frac{1}{8}$ of the cabins on Suncatcher are standard cabins. $\frac{1}{8}$ of 600 = 75

Blessy is incorrect.

There are more standard cabins on Suncatcher.

We need to remember that the two pie charts do not represent the same total number. The pie chart for Starlight represents 120 cabins and the pie chart for Suncatcher represents 600 cabins.

We can use the approximate fractions of the whole and the total number of cabins represented on each pie chart to check the claim.

Worked example 2

A ticket office takes bookings for a concert by telephone.

The frequency diagrams show how long (in minutes) customers phoning the ticket office have to wait before their call is answered.

The times are shown for calls received in the morning and the afternoon on one day.

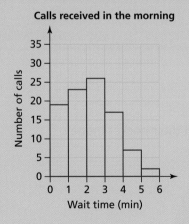

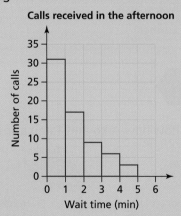

Work out the percentage of all the calls made that day that were answered within 2 minutes.

Morning Total number of calls = 19 + 23 + 26 + 17 + 7 + 2 = 94 Number of calls answered within 2 minutes = 19 + 23 = 42	First find the total number of calls made in the morning and the number of calls answered within 2 minutes.	**Calls received in the morning**
Afternoon Total number of calls = 31 + 17 + 9 + 6 + 3 = 66 Number of calls answered within 2 minutes = 31 + 17 = 48	Then find the total number of calls made in the afternoon and the number of calls answered within 2 minutes.	**Calls received in the afternoon**
Whole day Total number of calls = 94 + 66 = 160 Number of calls answered within 2 minutes = 48 + 42 = 90 Percentage answered within 2 minutes = $\frac{90}{160} \times 100\% = 56\%$ (to nearest whole number)	Combine the numbers for the morning and the afternoon and express as a percentage.	90 calls answered within 2 minutes 160 calls in the whole day

Exercise 1

1 The pie charts show how Aiko and Tamika spend their pocket money.

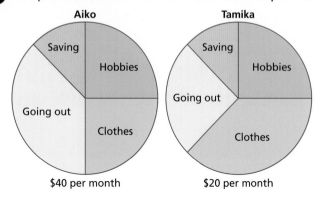

Aiko — Saving, Hobbies, Going out, Clothes — $40 per month

Tamika — Saving, Hobbies, Going out, Clothes — $20 per month

Write down if each of the conclusions below is true or false.

a) Tamika spends a greater proportion of her monthly pocket money on clothes than Aiko.

b) Aiko spends a smaller proportion of her monthly pocket money on going out than Tamika.

c) Aiko and Tamika save the same amount of money each month.

d) Aiko spends more money per month on hobbies than Tamika.

2 The pie charts show how land is used in India and in the UK.

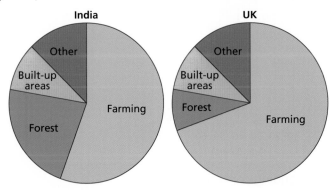

a) Write down three comparisons based on the pie charts.

b) What additional information would you need to be able to compare the amount of land that was used for farming in India and the UK?

3 The pie charts show how students travel to school at Alta Academy and Colham College.

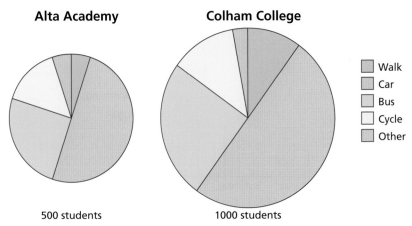

Legend: Walk, Car, Bus, Cycle, Other

500 students   1000 students

a) Write down the two methods of transport that were used by the same **proportion** of students at Alta Academy and at Colham College.

b) Which school has the larger **number** of students travelling to school by bus?

Show how you worked out your answer.

4 The diagrams show the ages of people attending two concerts.

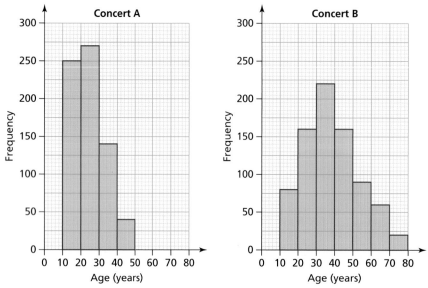

a) Write down the modal class for people attending concert A.

b) Write down the number of people aged over 70 years who attended concert B.

c) Find how many more people aged under 30 years attended concert A than concert B.

d) How do the ages of the people attending the two concerts compare?

5 The bar chart shows the number of people who attend the local gym before 9 a.m. every morning.

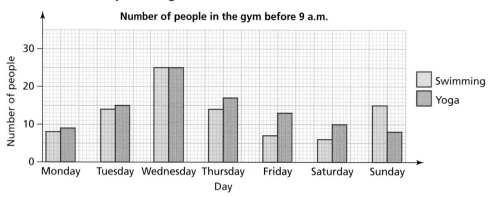

Which day had the highest proportion of people doing yoga?

Show how you worked out your answer.

6 The pie charts show the favourite type of movie for some children and some adults.

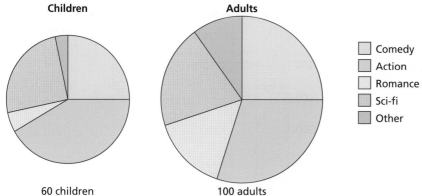

Children

Adults

	Comedy
	Action
	Romance
	Sci-fi
	Other

60 children   100 adults

a) Compare the proportion of children and the proportion of adults who said their favourite type of movie was romance.

b) Compare the proportion of children and the proportion of adults who said their favourite type of movie was action.

c) Compare the number of children who said comedy was their favourite type of movie and the number of adults who said comedy was their favourite type of movie.

d) Give a possible reason why the pie chart for the adults has been drawn with a larger radius than the pie chart for children.

7 Angelique has a collection of ancient Roman and Greek coins.

The table shows the diameters of the Roman coins in her collection.

Diameter, d (cm)	$1 < d \le 1.5$	$1.5 < d \le 2$	$2 < d \le 2.5$	$2.5 < d \le 3$	$3 < d \le 3.5$
Frequency	6	8	12	7	1

a) Draw a frequency diagram to show the diameters of Angelique's Roman coins.

The frequency diagram below shows the diameters of her Greek coins.

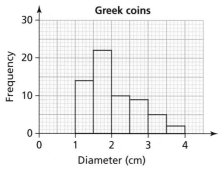

Greek coins

b) How many more Greek coins does Angelique have than Roman coins?

c) Calculate the percentage of all Angelique's coins that have diameter greater than 2.5 cm.

8 Elise is investigating changes in the population of Sweden between 1965 and 2016.

The frequency diagrams show the ages of people living in Sweden in these two years.

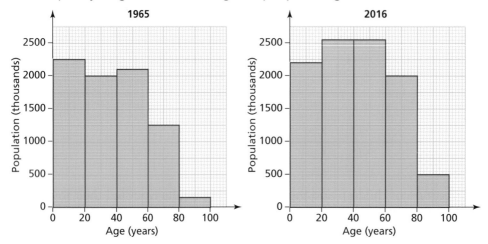

Elise concludes that the number of people aged 60 and over in Sweden has doubled between 1965 and 2016. Comment on Elise's conclusion.

Thinking and working mathematically activity

Real data question Investigate the change in land use in Portugal and New Zealand between 1990 and 2016.

Land use in Portugal

Type of land	Area in 1990 (km²)	Area in 2016 (km²)
Arable land and permanent crops	31 250	17 383
Permanent meadows and pastures	8380	18 758
Forest	34 360	31 706
Other areas	17 510	23 759

Land use in New Zealand

Type of land	Area in 1990 (km²)	Area in 2016 (km²)
Arable land and permanent crops	26 920	6450
Permanent meadows and pastures	134 900	100 060
Forest	96 580	101 522
Other areas	4910	55 278

Source: [© FAO] 2020 FAOSTAT Land Use data http://www.fao.org/faostat/en/#data/RL 15/09/2020

- Draw suitable diagrams to show the proportion of land area that is given over to different types of land. Give a reason for your choice of graph.
- Make conclusions from your graphs.
- Investigate land use in a country of your choice. Display your results statistically.

Did you know?

35% of land area in the world is used for agriculture.

Key terms

Averages are used to decide whether the values in one set of data are typically larger or smaller than the values in a second set of data. The measures of average are the **mean**, the **median** and the **mode**.

The **range** is a measure of **spread**. It is used to compare variations within the values of a data set.

A set of data is more consistent than a second set if its values are less spread out.

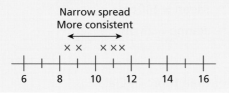

Discuss

The range can sometimes be an unreliable measure of the variability of a set of data.
Why might this be?

Worked example 3

A team plays football matches on Wednesdays and Saturdays.

The table summarises the attendances at 16 Wednesday matches and 16 Saturday matches.

	Wednesday	Saturday
Median	17 250	31 418
Mean	19 315	34 192
Range	14 436	7515

a) Compare the average attendances at the matches on the two days.

b) On which day were attendances more varied? Use the data in the table to explain your choice of answer.

a) The median and mean attendance is higher on Saturday. So on average attendances at the matches are higher on Saturday.	The median and the mean are both measures of average. Compare the values of these for the two days.	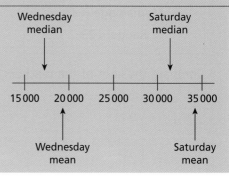

| **b)** The range is greater on Wednesday, so attendances are more varied on Wednesday. | The range measures how varied the data are. Compare the value of the range for Wednesday with the value for Saturday. | 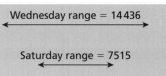 Wednesday range = 14 436 Saturday range = 7515 |

1 Billy is comparing the heights of the students in two classes.

Class A	Class B
Mean = 143 cm	Mean = 140 cm
Range = 52 cm	Range = 28 cm

Complete these statements.

a) On average, students in Class _____ are taller than those in Class _____ because the mean is higher.

b) The heights of students in Class _____ are more consistent than those in Class _____ because the range is lower.

2 Antonia is comparing the salaries of men and women in a company.

Men	Women
Mean = $28 000	Mean = $23 000
Range = $42 000	Range = $44 000

Decide if each statement below is true, false or you cannot tell.
Give reasons for your answers.

a) On average, the salaries of women are higher than men.

b) The salaries for men are more consistent than those for women.

c) The highest paid employee is a woman.

3 Here is some information about the ages of the teachers at two schools:

Valley High School	Greendale Academy
Median age = 35	Median age = 44
Range = 23	Range = 39

Compare the ages of the teachers in the two schools.

4 Here are the scores of two diving teams in a competition:

Team 1	9	10	7	8	7	8	7
Team 2	5	10	10	10	10	5	6

a) Find the mean score for each team.

b) Find the range of scores for each team.

c) Compare the performance of the two teams.

5 Two runners compete in a 100 m race.

Here are their performance statistics for a whole season:

Runner 1
Mean time = 10.7 seconds
Range of times = 1.2 seconds

Runner 2
Median time = 10.5 seconds
Range of times = 0.8 seconds

Megan says, 'Runner 1 has done better because he has a higher average time than Runner 2'. Do you agree with Megan? Explain your answer.

6 Two football teams compare the number of goals they have scored this season.

Team A
Mean number of goals scored = 2.1
Range of goals scored = 3

Team B
Mean number of goals scored = 1.9
Range of goals scored = 5

Chen says, 'Team B has done better because they have a higher range of goals than Team A'. Do you agree with Chen? Explain your answer.

7 The table shows the ages of people watching two films.

Age, x years	Frequency for Film A	Frequency for Film B
$10 \leq x < 20$	13	5
$20 \leq x < 30$	67	11
$30 \leq x < 40$	29	16
$40 \leq x < 50$	7	47
$50 \leq x < 60$	1	55
$60 \leq x < 70$	3	23

a) Write down the modal class interval for the ages of people watching Film A.

b) Write down the modal class interval for the ages of people watching Film B.

c) Compare the ages of people watching the two films.

8 A class takes tests in Geography and Science.

The marks in the Geography test are shown in the stem-and-leaf diagram.

Stem	Leaf
4	5
5	0 2 6
6	4 5 6 6 8 9
7	0 1 4 7 8
8	
9	2

Key 4 | 5 represents 45%

The mean and range of the marks in the Science test were:

mean = 68%   range = 31%

Compare the marks the class got in the two tests. Show your working.

9 Fabian records the numbers of words in 15 books written by each of two authors. The table summarises his results.

	Author A	Author B
Median	38 136 words	43 815 words
Mean	41 587 words	44 109 words
Range	17 397 words	12 765 words

a) Compare the average number of words in the books written by the two authors.

b) Which author writes books that are more consistent in length?
   Use the data in the table to explain your answer.

10 The table summarises the ages of people attending two concerts.

	Concert X	Concert Y
Median age	38 years	37 years
Mean age	38.4 years	39.2 years
Range of ages	64 years	48 years

a) Compare the spread of ages of the people attending the two concerts.

b) Amy wants to know which concert had younger people attending on average.
   What difficulty will Amy have in making a conclusion?

Thinking and working mathematically activity

Real data question Investigate how times in the Olympics 800 metres final have changed between 1964 and 2016.

Women (seconds)	
1964	**2016**
121.1	115.3
121.9	116.5
122.8	116.9
123.5	117.0
123.9	117.4
125.1	117.7
125.8	119.1
125.8	119.6

Men (seconds)	
1964	**2016**
105.1	102.2
105.6	102.6
105.9	102.9
105.9	103.4
106.6	103.6
107.0	104.2
107.2	106.0
110.5	106.2

Source of data: www.olympic.org

You should:

- decide on exactly what you are going to investigate and make some predictions
- calculate suitable summary measures and give a reason for your choice of measures
- make some conclusions and relate these back to your original predictions
- try to give some reasons for any differences you have discovered
- make some suggestions about how the investigation could be extended or improved.

24.3 Project

Thinking and working mathematically activity

Real data question Look at the data provided.

Europe

Country	1975–1980	1990–1995	2005–2010	2015–2020
Albania	3.90	2.79	1.60	1.78
Austria	1.65	1.48	1.40	1.53
Belgium	1.70	1.61	1.82	1.83
Croatia	1.90	1.52	1.52	1.48
Denmark	1.68	1.75	1.85	1.73
Finland	1.66	1.82	1.84	1.77
France	1.86	1.71	1.97	1.99
Greece	2.32	1.37	1.46	1.30
Hungary	2.13	1.74	1.33	1.40
Italy	1.89	1.27	1.42	1.49
Lithuania	2.10	1.82	1.42	1.63
Malta	2.12	1.99	1.39	1.49
Netherlands	1.60	1.58	1.75	1.77
Norway	1.81	1.89	1.92	1.81
Portugal	2.55	1.48	1.37	1.24
Slovakia	2.46	1.87	1.31	1.44

Africa

Country	1975–1980	1990–1995	2005–2010	2015–2020
Angola	7.35	7.15	6.60	5.79
Benin	7.00	6.56	5.31	4.50
Botswana	6.37	4.32	2.90	2.57
Cameroon	6.47	6.22	5.21	4.46
Côte d'Ivoire	7.81	6.41	5.36	4.77
Eritrea	6.62	6.20	4.80	4.02
Ethiopia	7.18	7.09	5.26	3.99
Gambia	6.34	6.08	5.79	5.53
Ghana	6.69	5.34	4.29	3.95
Guinea	6.45	6.51	5.54	4.73
Kenya	7.64	5.57	4.80	4.10
Lesotho	5.69	4.70	3.37	3.01
Morocco	5.90	3.70	2.49	2.38
Namibia	6.60	4.91	3.60	3.31
Sierra Leone	6.25	6.62	5.51	4.28
Tunisia	5.65	2.98	2.02	2.07

Source: 'World Population Prospects: Key findings & advance tables 2015 revision, by the Department of Economic & Social affairs, © 2015 United Nations. Used with the permission of the United Nations.

The tables give information about the average number of children born to women in 16 European and 16 African countries. Data have been given for four different time periods.

* Decide on a hypothesis that you could investigate using some or all of the data. Your hypothesis should involve some comparisons.

* Draw some relevant graphs.

* Summarise your data by calculating averages and the range.

* Form some conclusions.

* What could you do to extend your investigation? Perhaps you could find some secondary data yourself that you could investigate.

* Think about the limitations of your work. What could you have done to make your work better?

Consolidation exercise

1 The pie charts show the types of chocolates in two different chocolate selections.

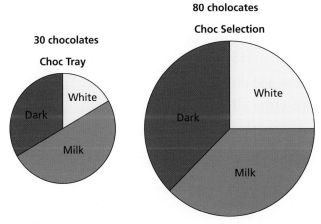

80 cholocates

Choc Selection

30 chocolates

Choc Tray

a) Obafemi says that there are more milk chocolates in the Choc Tray box than in the Choc Selection box. Is Obafemi correct? Explain your answer.

b) Write down a comparison of the most common type of chocolate in the two boxes.

c) Write down a comparison of the least common type of chocolate in the two boxes.

d) Suggest a reason for the Choc Tray pie chart being drawn with a smaller radius than the Choc Selection pie chart.

2 The frequency diagrams show the temperatures in two cities during June and July.

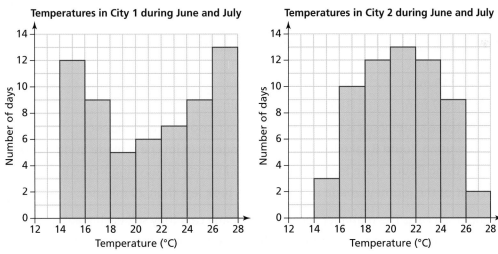

Serena wants to visit one of the cities during June or July next year.

She wants to choose the city where the temperature is fairly predictable.

She doesn't like it too hot or too cold.

Which city would you suggest Serena visits? Explain your answer.

3 Erin has collected the pulse rates of a group of adults before and after exercise, measured in beats per minute. The results are shown in the lists below:

Before exercise

59	61	65	68	68	69	70	71	71	74
76	76	77	78	83	86	87	88	91	95

After exercise

87	91	93	93	95	98	98	99	100	101
101	103	106	110	111	112	115	118	120	121

a) Calculate the mean pulse rate before exercise and after exercise.

b) Describe what the data shows about how exercise affects pulse rate.

4 The table shows some information about the ages of three groups of people.

	Group 1	Group 2	Group 3
Mean	17 years	19 years	18 years
Minimum	3 years		
Maximum	34 years		
Range		31 years	31 years

a) Compare the average age of the people in the three groups.

b) Sally says, "The range for all three groups is the same, so the maximum age of the people in all three groups must be the same."

Comment on Sally's conclusion.

5 Dom can take two different routes to work. He records how long (in minutes) it takes him to travel to work by each route on 20 different occasions.

His journey times are summarised below.

Route A	Route B
Mean time: 18 min	Mean time: 16 min
Median time: 19 min	Median time: 18 min
Range of times: 7 min	Range of times: 4 min

Decide if each of these statements is true or false. Give a reason for each answer.

a) Journeys made using Route B take less time on average than journeys made using Route A.

b) Journey times using Route A are less varied than the journey times using Route B.

c) More than half of the journeys made using Route A took less than 20 minutes.

6 Henri and Claudia sell cars. The manager records how many cars they sell each week.

The table shows the mode, mean and range for the number of cars they sold each week over the past year.

	Henri	Claudia
Mode	4 cars	3 cars
Mean	3.8 cars	3.9 cars
Range	5 cars	9 cars

The manager wants to give a bonus to either Henri or Claudia.

a) Henri thinks that he should be given the bonus. Use the numbers in the table to explain why he might think that.

b) The manager actually gives the bonus to Claudia. How can the manager justify his decision?

End of chapter reflection

You should know that...	You should be able to...	Such as...
Distributions can be compared by examining the shape of their graphs.	Compare two or more sets of data and relate conclusions to the original question.	The pie charts show the sciences studied by students at two colleges. Make a comparison of the subjects studied.
You can compare two sets of data by comparing their average values (mean, median or mode) and their spread (range). You say that the data are more consistent when they have a lower range and the values are closer together.	Compare two (or more) sets of data using the range, mean, median and mode.	The table summarises the daily temperatures in two cities in June. <table><tr><td></td><td>City A</td><td>City B</td></tr><tr><td>Mean</td><td>19.3 °C</td><td>17.7 °C</td></tr><tr><td>Median</td><td>19 °C</td><td>16 °C</td></tr><tr><td>Range</td><td>5.5 °C</td><td>8 °C</td></tr></table> Compare the temperatures of the two cities.

Accurate drawing

You will learn how to:

- Construct triangles, midpoint and perpendicular bisector of a line segment, and the bisector of an angle.
- Represent front, side and top view of 3D shapes to scale.

Starting point

Do you remember…

- how to use a ruler, protractor and set square to construct a quadrilateral?

  For example, construct this quadrilateral.

  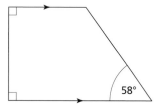

- how to use ratio notation?

  For example, a paint is made by mixing blue and yellow paint in the ratio 1 : 3

  If I use 2 litres of blue paint, how much yellow paint do I use?

- how to draw representations of 3D shapes?

  For example, draw a plan, front elevation and side elevation of this cuboid.

  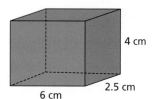

This will also be helpful when…

- you learn to solve problems about loci and bearings
- you use maps in real life
- you find the surface area of 3D shapes.

Mystic Rose

Begin by drawing a circle.

Next mark six equally spaced points around the edge of the circle (the easiest way to do this is to leave your compass radii the same size and use each point as the centre to mark the next point).

Next draw straight lines between each point and the others.

Keep going until you have joined all of the points.
This is a Mystic Rose.

- Can you identify and name some of the shapes you can see in the Mystic Rose?
- Can you find how many lines there are altogether?
- Now investigate with different numbers of points around the edge.
- Can you find a link between the number of points around the edge and the number of lines in the Mystic Rose?

25.1 Construction of triangles

> ### Did you know?
>
> The triangle is the only polygon that cannot be deformed without changing the lengths of one of its sides.
>
> For example, a square is easily deformed to make a rhombus and a rectangle is easily deformed to make a parallelogram.
>
> This is why triangles are used a lot in construction.
>
> Use the internet to find other images of triangles being used in construction.
>
>

Construct means to draw accurately. You will usually need to use equipment such as a ruler, protractor and a pair of compasses to produce accurate drawings.

You can construct a triangle when you know sufficient information.

SAS is used as shorthand for 'given two sides and the included angle' – side, angle, side.

ASA is used as shorthand for 'given two angles and the included side' – angle, side, angle.

SSS is used as shorthand for 'given three sides' – side, side, side.

RHS is used as shorthand for 'given a right angle, hypotenuse and one side' – right angle, hypotenuse, side.

The minimum information needed to construct a triangle is either – SAS, ASA, SSS or RHS.

For example,

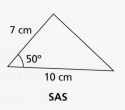

SAS

ASA

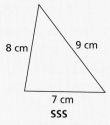

SSS

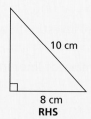

RHS

Worked example 1 (ASA and SAS constructions)

Construct these triangles.

a)

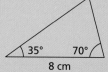

b)

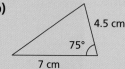

a) Use a ruler to draw an 8 cm line segment. This will be the base of the triangle.
   Now construct a 35° angle at the left end.

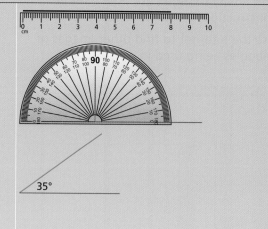

Now position the protractor on the right end of the base line.

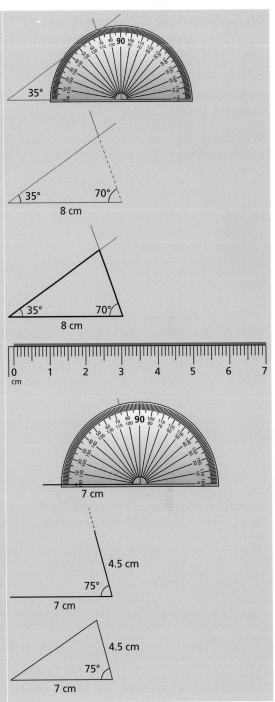

Construct a 70° angle at the right end of the base line.

Extend the new line as needed to cross the existing (green) line.

You now have your triangle.

b) Use a ruler to draw a 7 cm line segment. This will be the base of the triangle.

Position the protractor on the right end of the base line. Align the 0° line so it is exactly over the base line.

Reading round from 0°, make a mark at 75°.

Remove the protractor and connect the right end of the base to the mark you just made. Measure 4.5 along this line.

Join the ends of the 7 cm and 4.5 cm lines to make a triangle.

Worked example 2 (SSS construction)

Use a ruler and a pair of compasses to construct a triangle with side lengths of 7 cm, 4 cm and 5 cm.

Start by drawing a 7 cm line using your ruler.

A ————————— B
7 cm

Remember to write the lengths on your sides as you draw the triangle.

Now open your pair of compasses so that the distance between the point and the tip of the pencil is 4 cm.

Put the point of your pair of compasses at *A* and draw an arc. Make sure that the pair of compasses stays opened to 4 cm.

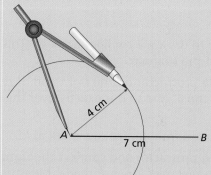

Now open your pair of compasses to 5 cm and draw an arc with the point of the compasses at point *B*.

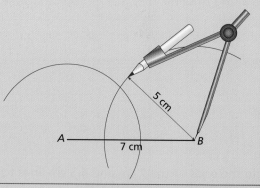

Join the ends of the base line to one of the points where the arcs cross.

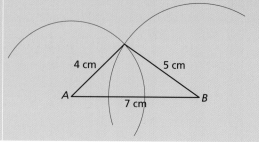

Worked example 3 (RHS construction)

Use a ruler and a pair of compasses to construct a triangle with a right angle, a hypotenuse of length 9 cm and a side of length 6 cm.

Start by drawing a straight line segment and make a mark part way along it.	

The next step is to draw a right angle at the mark.

Open your pair of compasses and use them to make arcs crossing the line segment on either side of the mark.

Open your pair of compasses further and draw an arc from each of points *A* and *B*.

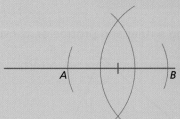
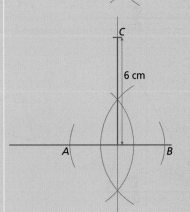

Draw a straight line segment through the two points where the arcs cross.

The angle formed with the horizontal line is 90°.

Measure 6 cm along the vertical line and make a mark, labelling this point *C*.

Open your pair of compasses to 9 cm (the length of the hypotenuse). Draw an arc from the point *C*.

Join the point *C* to the point where the 9 cm arc crosses the horizontal line.

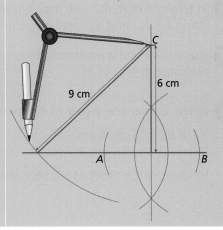

If you are given three sides of a triangle, is there only one possible triangle?

In a RHS construction, is there only one possible triangle?

If you were given three angles of a triangle, is there only one possible triangle?

Exercise 1

1 Using a ruler and protractor, construct these triangles.

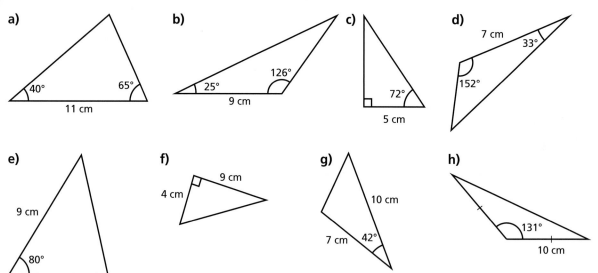

a) 40° 65° 11 cm

b) 25° 126° 9 cm

c) 72° 5 cm

d) 7 cm 33° 152°

e) 9 cm 80° 6 cm

f) 9 cm 4 cm

g) 10 cm 7 cm 42°

h) 131° 10 cm

2 In triangle *PQR*, *PQ* = 60 mm, angle *PQR* = 48° and angle *RPQ* = 61°.
Using a ruler and a protractor, construct an accurate drawing of triangle *PQR*.

3 Use a ruler and a protractor to construct an isosceles triangle with base 4.4 cm and base angles of 35°.

4 a) Use a ruler and a pair of compasses to construct a triangle with sides measuring

    i) 8 cm, 6 cm, 5 cm   **ii)** 3 cm, 4 cm, 5 cm   **iii)** 6 cm, 8 cm, 6 cm
    iv) 7.5 cm, 6.3 cm, 4.7 cm   **v)** 7.3 cm, 7.3 cm, 7.3 cm   **vi)** 5.4 cm, 8.5 cm, 4.2 cm

b) Identify the types of triangle drawn in part **a)**.

5 Use a ruler and a pair of compasses to construct a triangle ABC with

   a) $ABC = 90°$, $AB = 6$ cm, $AC = 10$ cm   **b)** $ABC = 90°$, $AB = 4.5$ cm, $AC = 7.8$ cm

6 Ivan is constructing an SSS triangle with side lengths 6 cm, 5.5 cm and 7.3 cm.
   Here is his diagram.

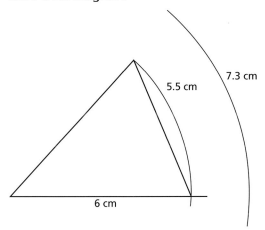

6 cm

5.5 cm

7.3 cm

   Identify the mistakes in Ivan's work and draw an accurate version of the triangle.

7 Use a ruler and a protractor to construct an isosceles triangle with base 5 cm and the base angles equal to twice the other angle.

8 Hua is asked to draw a triangle ABC with $AB = 8$ cm, $AC = 5$ cm and angle $ABC = 30°$.
   He says there are two possible different triangles.
   Show Hua is correct.

Thinking and working mathematically activity

Technology question Roll a ten-sided dice three times, or just write down three random numbers less than 10.

• Use dynamic geometry software to construct a triangle with side lengths in centimetres equal to your three numbers.

• Repeat this with three different numbers.

• What combination of numbers can you not use to draw a triangle?

• Try to find a rule to check whether three side lengths will give a triangle without having to try drawing it.

25.2 Constructing bisectors

Key terms

The **midpoint** of a line segment is the point that is the same distance from both end points. It is halfway along the line segment.

A **line bisector** is a line which cuts a line in half.

An **angle bisector** is a line that cuts an angle in half.

Worked example 4

a) Using a pair of compasses and a ruler, construct the perpendicular bisector of an 8 cm line.

b) Construct the angle bisector of a 70° angle.

a) First draw an 8 cm line segment.

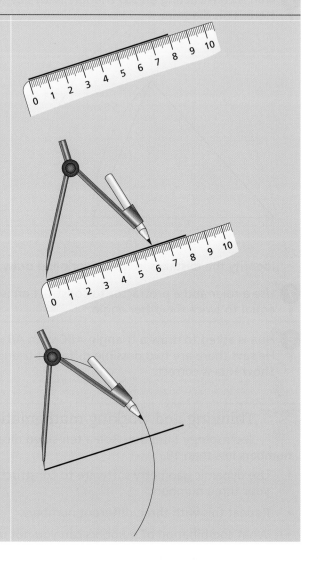

Set your pair of compasses so that the point and the pencil are more than half of the length of your 8 cm line segment.

Place the point on one end of the line segment and draw a large arc.

Do not adjust your pair of compasses. Now place the point on the other end of the line segment and again draw a large arc. The two arcs should cross twice. If they do not cross twice, then extend your arcs.

Draw a straight line through the two points where your arcs meet. This is the **perpendicular bisector**.

Do not rub out your construction lines.

This is the midpoint of the line segment

b) Use your protractor to draw an angle of 70°.

Place the pair of compasses on point *A* and draw an arc.

A

Now place the pair of compasses on point *B* and draw another arc. Do not alter the settings of your pair of compasses.

Now place the pair of compasses on the point *C* and draw a final arc.

Finally draw a straight line through the points where your arcs meet and point *A*.
This is the **angle bisector**.

Do not rub out your construction lines.

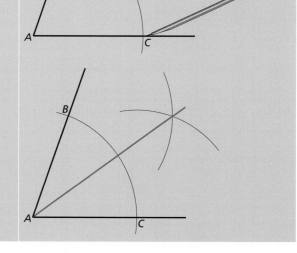

Think about

What size angles could you construct by combining the two techniques from Worked example 4?

Exercise 2

1 Draw a line segment measuring 12 cm. Construct the perpendicular bisector.

2 Draw a line segment measuring 9 cm. Find the midpoint of the line by constructing the perpendicular bisector.

3 Use a protractor to draw a 50° angle. Construct the angle bisector.
Use your protractor to check that you have done this accurately.

4 Use a protractor to draw a 130° angle. Construct the angle bisector.

5 Draw a line that is 10 cm long.
   a) Construct the perpendicular bisector of the line.
   b) Bisect one of the angles between the original line and the perpendicular bisector.

6 **a)** Draw a triangle. Make it fairly large so that your constructions are accurate.
   b) Construct the perpendicular bisector of each side. Write down what you notice.
   c) Investigate with other starting triangles.
   d) Find a triangle where the perpendicular bisectors of the sides meet outside the triangle.

▼ Thinking and working mathematically activity

Use dynamic geometry software to draw a triangle and extend two of the sides.

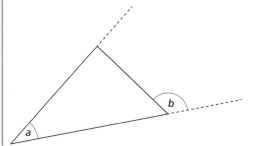

Construct bisectors of angles *a* and *b*.

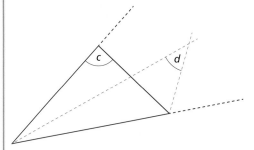

Now measure the angles *c* and *d* on your triangle.

What do you notice about angles *c* and *d*?

Now draw a different triangle, bisect the angles and measure as before. Is the relationship the same?

What about if you use special types of triangles, such as equilateral, obtuse or isosceles triangles? can you explain why this relationship works?

Worked example 5

Draw scale drawings of the plan and elevations of this garden shed. Use a scale of 1 : 200

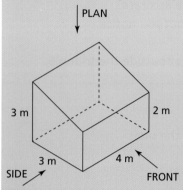

Plan	The scale 1 : 200 means that 1 cm on the plan represents 200 cm = 2 m in real life.
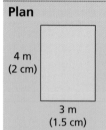	The plan view shows the view from the top looking down. The 3 m side will be represented by 3 ÷ 2 = 1.5 cm on the drawing. The 4 m side will be represented by 2 cm on the drawing. Mark the measurements on the drawing.
Side elevation	The side elevation shows the view of the shed from the side.
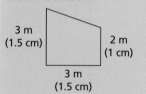	
Front elevation	The front elevation shows the view of the shed from the front.
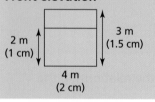	

Exercise 3

1 Draw the plan, side and front elevations of this rectangular solid. Use a scale of 1 cm to 2 m.

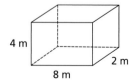

2 Draw the plan, side and front elevations of this triangular prism. Use a scale of 1 : 5

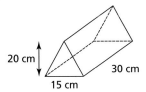

20 cm

30 cm

15 cm

3 Draw the plan, side and front elevations of this cylinder. Use a scale of 1 : 2

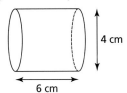

4 cm

6 cm

4 Draw the plan, side and front elevations of this structure. Use a scale of 1 cm : 5 m.

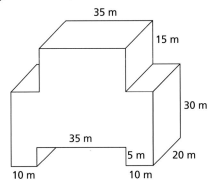

35 m

15 m

30 m

35 m

5 m   20 m

10 m   10 m

5 Draw the plan, side and front elevations of these shapes. Use the scale shown.

a)

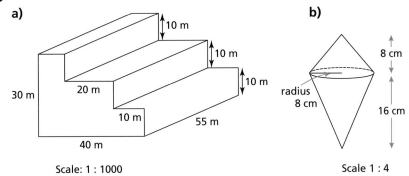

10 m

10 m

10 m

30 m

20 m

10 m

55 m

40 m

Scale: 1 : 1000

b)

8 cm

radius
8 cm

16 cm

Scale 1 : 4

6 Sophia drew a plan and elevations of this building.

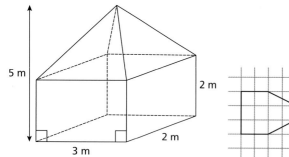

5 m

2 m

2 m

3 m

She has made some errors.
Explain what Sophia has done wrong and correct her errors.

Thinking and working mathematically activity

- These diagrams show the plan, side elevation and front elevation of three objects.

a)

b)

c)
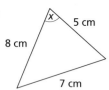 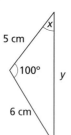

- Describe each object and sketch a diagram to illustrate your description.
- Create three images of your own for someone else to guess.

Consolidation exercise

1 Construct the following triangles and measure the lengths or angles indicated by a letter.

a)
x
5 cm
8 cm
7 cm

b)
5 cm
x
100°
y
6 cm

c)
x
y
6 cm
10 cm

2 In triangle *ABC*, *AB* = 5.2 cm, *AC* = 6.7 cm, angle *BAC* = 58°
a) Construct an accurate drawing of triangle *ABC*.
b) Measure the length *BC* to the nearest mm.

3 Construct an isosceles triangle with base 7 cm and base angles of 55°
Measure the length of one of the other sides.

4 Draw a 10 cm line segment on plain paper. Find the midpoint of the line segment
by constructing the perpendicular bisector.

---

5 Use a protractor to draw a 70° angle. Construct the angle bisector.
Use your protractor to check that you have done this accurately.

6 Katrina draws a triangle. The lengths of two sides are 5 cm and 3 cm.
The angle between these sides is 53°.
She says, 'It's a right-angled triangle.'
Construct a diagram to decide if Katrina is correct.

7 Draw the plan, side and front elevations of the buiding below. Use a scale of 1 cm to 2 m.

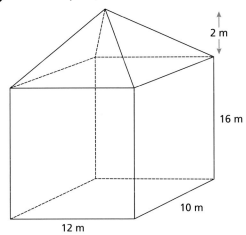

2 m

16 m

10 m

12 m

End of chapter reflection

You should know that...	You should be able to...	Such as...
SSS, SAS, ASA and RHS are minimum requirements to draw a triangle.	Construct a triangle given the minimum information.	Draw the triangle *ABC* where *AB* = 6 cm, *BC* = 5 cm and angle *ABC* = 73°
The midpoint is halfway along a line segment. A bisector cuts something in half.	Construct a midpoint and perpendicular bisector of a line segment. Construct the bisector of an angle.	Draw a line segment 6 cm long and construct its perpendicular bisector. Draw an angle of 44° and construct its angle bisector.
A plan is a view from above. An elevation is a view from the side or the front.	Represent plan, side and front elevations of 3D shapes to scale.	Draw the plan, side elevation and front elevation for this prism, with a scale of 1 : 25 75 cm 2 m 1 m